Environmental Controls

Environmental Controls

The Impact on Industry

Edited by
Robert A. Leone
Harvard University

Lexington Books
D.C. Heath and Company
Lexington, Massachusetts
Toronto   London

Library of Congress Cataloging in Publication Data

Main entry under title:

Environmental controls.

Includes index.
1. Water quality management—United States—Costs. 2. Costs,
Industrial—United States. 3. Environmental policy—United States.
I. Leone, Robert A.
TD223.E58   658.1'553   75-32222
ISBN 0-669-00345-x

Published simultaneously in Canada

Printed in the United States of America

International Standard Book Number: 0-669-00345-x

Library of Congress Catalog Card Number: 75-32222

Contents

List of Figures

List of Tables

Preface

This volume builds upon a sixteen-month study of the impact on manufacturing industries of the 1972 amendments to the Federal Water Pollution Control Act. The initial research was sponsored by the National Commission on Water Quality[a] and carried out at the National Bureau of Economic Research.[b] In addition to summarizing many of our research findings presented to the National Commission, we have attempted to identify and emphasize the important generalizations that we feel emerged in the parallel study of the impact of federal water pollution control requirements on six different industries by six different sets of authors. In essence, therefore, this volume is our attempt to take a retrospective look at our own research done under the time pressures of a research contract.

Our study is aimed simultaneously at those persons responsible for the execution of economic impact studies and those persons who are the subjects of such studies.

We have attempted to stress the methodology of impact analysis and in doing so have attempted to be critical of our own practices. Hopefully, future researchers will be able to avoid some of the problems we faced. It is also our hope that individuals in industry, when called upon to critique an impact study, will see from our analysis what factors we feel a good study ought to address with particular scrutiny and what factors can be treated in a less rigorous fashion.

Although this volume will also be useful to individuals concerned with the specific empirical findings of our research, we strongly recommend that the reader interested in a more detailed treatment of any individual industry refer directly to the various reports to the Commission upon which we have so heavily drawn.

As the list of authors of various chapters indicates, a large number of individuals were involved in the preparation of this volume and the predecessor reports to the Commission. Several of the contributors provided much greater input than the authorship of an individual chapter or two might suggest. Dick Startz and Jim Smith both were more generous of their time and effort than any editor should reasonably expect. Monty Blanchard

---

[a] This volume is based upon research carried out under contract (#WQ4ACO14) to the National Commission on Water Quality. The Commission, however, bears no responsibility for either the findings or the interpretations of the findings reported here.

[b] Although this volume is based upon research done under the auspices of the National Bureau of Economic Research, the findings reported herein have not undergone the full critical review accorded National Bureau studies, including review by the Board of Directors, and therefore the NBER bears no responsibility for the findings of this analysis.

read and reread many drafts of various chapters in addition to the one he coauthored.

The authors have greatly profited from the personal assistance of many individuals in the preparation of this work, particularly Dale Whittington of the National Commission staff.

As editor, I would also like to express my personal thanks to Jim McKenney, Paul Olsen, Bob Hayes, Bob Stobaugh, and Paul Marshall of the Harvard Business School faculty for their comments on various drafts of the material included in this report.

John Meyer and Harvey McMains, Bill Apgar, and John Kain of the National Bureau also assisted us at crucial times during the research, providing much needed encouragement as well as constructive advice.

Rose Giacobbe and her staff at Harvard Business School provided very able clerical assistance in preparing this manuscript. Rachael Daitch helped to expedite the manuscript's production. We are particularly indebted to Mary Ellen Patrell who made many useful editorial suggestions.

The work has also benefited from generous institutional support. Although the National Commission provided the initial financial support, the Division of Research at the Harvard Business School, through the generosity of the School's Associates, provided the resources which permitted us to continue our research after the contract expiration date so that we might actually produce this volume.

I should also note that initial encouragement for us to undertake this project came from John Myers of the Conference Board. His enthusiasm for the project coupled with the enouragement of Joe Moore, our Project Director at the National Commission, determined both the scale of the project and the initial direction of our research.

In thanking all those who have helped in this project, I must note that the errors of analysis which remain in this volume are entirely the responsibility of the authors.

Robert A. Leone, Editor

Introduction

During the last several years an incredible array of environmental protection legislation has passed through Congress and the various state legislatures. Although the government had long been involved in environmental protection, increased popular awareness of environmental problems in the 1970s focused attention on the government's policies in this area.

To persuade constituents of their concern, legislators passed pervasive and highly restrictive controls. To convince the business community of their resoluteness, the legislators attached high penalties to noncompliance and often dictated less than generous compliance schedules.

The 1972 Amendments to the Federal Water Pollution Control Act (the original bill was passed in 1956) are typical of this legislation. The impact of these amendments on several manufacturing industries—which is the specific focus of this book—is only now being felt.

This particular piece of legislation not only significantly tightened existing performance criteria for effluent controls in manufacturing, but also mandated a rigid compliance schedule. The legislation requires that by 1977 industries achieve water pollution control levels consistent with the "best practicable technology" (BPT) and by 1983 performance improve still further to reflect the "best available technology" (BAT). The Act goes so far as to establish as a goal the "end of discharge of pollutants" (EDOP) by 1985.

The initial reaction of many in the business community to this legislation was negative, because the new controls were considered prohibitively costly to implement. Lobbying efforts by trade associations to alleviate the more onerous burdens of compliance have proliferated. The courts are full of cases brought by impacted firms against the government.

Yet, for all the time and resources that both the federal government and private industry have devoted to the examination of the impact of this legislation on private enterprise, public understanding of this impact remains surprisingly unsophisticated. We report the results of our investigation of the nature of these impacts in the chapters that follow, in the hope that they will contribute to an improved understanding of these problems. Our study is hardly the "last word"; but we do hope that our results and the questions they raise will motivate government officials to rethink regulatory strategies, the business community to reassess its evaluation of the burden that environmental controls represent, and researchers to reevaluate their research strategies.

We should also note that our principal motivation does not stem solely from a concern for the specific consequences of this isolated piece of legislation. Rather, we see the imposition of water pollution controls as an increasingly common exercise in the federal regulation of industry.

Abstracting from the merits of this legislation in terms of improvement in water quality—and we assume these benefits are substantial—the principal direct effect of legislated water pollution controls is to increase the costs of production in the affected industries. In recent years, much federal regulatory legislation has had precisely this effect. In addition to environmental controls, worker safety and some consumer protection regulation have had directly analogous impacts. In the past, one typically associated a "regulated industry" with the Interstate Commerce Commission, the Federal Communications Commission, or some similar agency which directly regulated entry, prices, and return on investment. Today, virtually every major industry in the "nonregulated sector" is subject to pervasive federal influences on its choice of production technologies, and the costs of utilizing those technologies. Water pollution controls are only one such influence.

The research findings reported here will contribute to an improved understanding of the impact of this growing form of government influence on the private decisions of business managers.

Choice of Methodology

There are several approaches one could take to the evaluation of the "impact" of a specific piece of legislation on industry. Perhaps the most obvious would be to perform a cost-benefit analysis to determine whether the benefits of improved water quality justify the higher production costs industry will incur to achieve these benefits. Such an approach is outside the scope of our analysis, for we do not attempt to place values on the improvements in environmental quality that will result from the enforcement of this legislation.

Nor do we choose to perform a cost-effectiveness analysis. We do not, for example, inquire as to whether tighter controls on, say, municipal sewage treatment facilities would achieve the desired improvements in environmental quality at lower cost than controls on, say, the pulp and paper industry.

Rather, we focus our attention somewhat narrowly on the consequences of this legislation for industry. Our focus is necessarily limited both because of the complexity of the issues and because of our view that one must first understand industry impacts *before* either a cost-effectiveness or cost-benefit analysis is possible.

Further, as we attempt to document in Chapter 1, these industry impacts cannot simply be equated with the higher production costs resulting from compliance with the effluent control legislation. We will attempt to demonstrate in Chapter 1 that the dominant effect of federal water pollution controls will be to change the cost structures of the various water-using industries. This will result in major consequences for industry capacity in the short run as firms spend the resources necessary to comply with the law; and will have competitive consequences in the long run, which may be the dominant consequence of environmental controls.

Organization of the Text

In Chapter 1 we discuss the methodology of economic impact analysis. What follows in Chapters 2 through 7 is an empirical application of our methodology to six different industries.

Perhaps the most straightforward conclusions are drawn in the petroleum refining industry (Chapter 2), where the relatively low cost of meeting the effluent guidelines combines with otherwise strong economic pressures in the industry to yield a set of reasonably robust conclusions.

In Chapter 3 the examination of the pulp and paper industry emphasizes short-run industry dynamics. Using a primitive simulation model of the industry, we trace the implications of diverting capital resources from investments in capacity to investments in pollution abatement equipment. We also examine possible product changes and process modifications that might lower the cost of compliance over time.

An analysis of the steel industry follows in Chapter 4. In this industry we examine the role that preferential federal, state, and local tax provisions play in reducing the cost of satisfying increasingly stringent water quality standards. We also briefly examine trends in technological change to illustrate how forces independent of pollution abatement technologies can determine the burden of environmental control on industry.

In Chapter 5 we examine the textile industry. As one of the most competitive industries included in our study, it serves to illustrate the possible geographic consequences of imposing pollution controls. In this chapter, the issue of plant closings due to imposition of pollution regulations is addressed in some detail, although briefer discussions of the same problem occur in other chapters as well.

Chapter 6 examines the aluminum industry, in which one company in particular faces high costs to comply with the Act. This industry illustrates very well the consequences of an uneven distribution of the costs of control within an industry.

Chapter 7 focuses on the metal finishing industry, where water pollution controls result in so pervasive a change in the industry's cost structure that methodology, based on principles of incremental analysis, is put to a severe test.

In the final chapter we outline the principal conclusions arising from our examination of the impact of water pollution controls on these six manufacturing industries.

Environmental Controls

1

The Methodology: A Framework for Analysis

*Robert A. Leone
and Richard Startz*

In undertaking industrial impact studies of the sort presented in this volume, one quickly discovers that there is no simple procedure that can be applied to all industries, nor indeed is there even a single "best" procedure for a given industry. Among the industries studied here, our investigative methodology varies according to the nature of the industrial environment and market structure, the magnitude of pollution control costs in comparison with other costs of operation, and the quality and quantity of relevant data. Nonetheless, underlying the different empirical procedures is a single framework for analysis.

Our methods represent no new addition to economic theory. We attempt, rather, to apply techniques of marginal economic analysis to questions of pollution control impact. Too many studies have failed to present an *economic* impact analysis, presenting, instead, a greatly detailed accounting of the cost of pollution regulations to *existing* manufacturing plants without recognizing that the very same regulations will change the way manufacturing is conducted. Other studies have presented involved econometric work depicting the markets for the products in industries experiencing controls, many times stressing the estimation of demand relationships. Such analyses often fail to make clear the economic assumptions that underlie the changes on the "supply side" of industrial markets which pollution controls may precipitate.

We will make our analytic framework explicit, and by so doing, will attempt to achieve three objectives:

First, by defining our methodology we hope to make clear our criteria for a comprehensive analysis of the economic impact on industry of federal water pollution controls.

Second, the statement of methodology makes clear the general assumptions we have made. For example, our analysis assumes that firms try to maximize profits. We feel that this is a good approximation to the way firms in the real world operate. Those who prefer to assume that firms operate so as to maximize social welfare or to achieve some other social objective will not consider our methodology an appropriate way to address problems of impact estimation. Others, who see firms as attempting to satisfy a complex

1

set of private economic objectives of which profit is only one component, will find our analysis simplistic, but adaptable to their needs.

Naturally, each individual industry study has its own specific assumptions as well. These are stated where appropriate in the chapters that follow.

The third purpose of this chapter is to permit the reader to compare how we would have conducted the studies if conditions had been perfect to how they were actually done in this less than perfect world. For example, our idealized analysis requires information as to the distribution of production costs among plants within each industry. This information was not available. We made assumptions and did without.

By specifying deviations from our ideal, we can often suggest the direction of the bias in our results caused by these omissions.

The Simple Model of Impact

A simple supply and demand diagram, the economist's most traditional tool, can be used to consider the issues we wish to examine. We find in Figure 1-1 four curves. S and S' are, respectively, the supply curves before and after the imposition of the pollution regulations we are examining (hereafter referred to as "before and after the Act"). The shape of these curves is drawn to reflect the fact that we are primarily interested in a short-run analysis, a time frame in which the intensity of manufacturing operations may be altered, but which is too short to allow substantial investment in new capacity. The supply curve is flat over most of the range, reflecting the notion that average unit costs are more or less constant so long as capacity is not strained. At the point of full capacity utilization, the curve turns to the vertical since it is not possible to get any more output at any cost.[a]

As S' illustrates, pollution regulations shift the supply schedule in two ways. First, they add to the cost of every unit produced.[b] In Figure 1-1 every point on the supply curve is raised by the amount $P'_1 - P_1$. Many studies stop at this point, taking the viewpoint that if costs rise by a given

[a] Of course, supply curves in the real world do not have this characteristic shape. Some savings can almost always be made as output drops, by retiring the least efficient equipment first and thus imparting a moderate slope to the horizontal portions of S and S'. Similarly, it is virtually always possible to squeeze a little more production out of existing facilities if only the reward is high enough, thus preventing the curves from ever becoming vertical. We have drawn the curves as we have, quite intentionally, however, to define extreme conditions. By constructing them in this manner, we know that the "true" S and S' must lie within these two bounds. In subsequent analysis, data limitations force us to rely on this fact to place boundaries on our impact estimates.

[b] This increase in cost is by no means uniform across producers, as Figure 1-1 suggests; this fact proves to be quite important to our subsequent analysis.

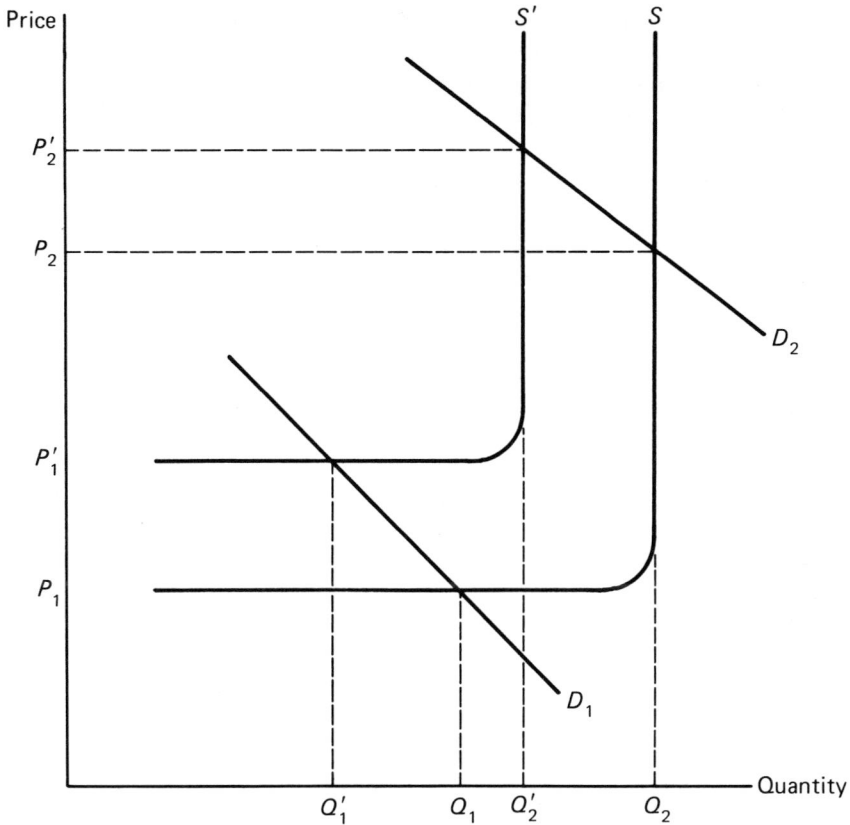

Notes: S, S': Supply before and after the Act, respectively.

D_1, D_2: Alternative specifications of demand.

Q_1, Q_2, Q_1', Q_2': Alternative equilibrium output levels before and after the Act, respectively.

P_1, P_2, P_1', P_2': Alternative equilibrium price levels before and after the Act.

Figure 1-1. Two Alternative Short-Run Equilibria

amount, so will prices. This is by no means an unreasonable view. In the long run, if the product is sold under conditions of free competition, if there is unrestricted entry into the market, if all factors of production trade freely, and if none are in fixed supply, then a cost increase ought to translate into an equivalent price increase. We think the key words in this set of conditions are "long run" and "free competition."

As the following discussion will attempt to demonstrate, during the process of competitive adjustment to a new long-run equilibrium, the short-run impact of pollution controls on price and output levels in an industry can substantially exceed the impacts that will occur over the long term. Similarly, we will attempt to show that the existence of noncompetitive forces can materially influence both the short-run and long-run impact of pollution controls on industry profits, prices, and output levels.

The issue of cost "pass-through" to consumers is typically a source of great debate. Initially, the size of the cost increase is debated. Subsequently, efforts are made to estimate the "demand elasticity" in the impacted industry. The demand elasticity indicates the percent change in consumption that will be associated with any given percentage change in industry price levels due to the added costs of effluent control. For example, an elasticity of −0.5 would indicate that a 1 percent *increase* in price would be associated with a 0.5 percent *decrease* in consumption; an elasticity of −1.5 would translate a 1 percent price increase into a 1.5 percent decrease in consumption. Clearly, the size of the demand elasticity, when coupled with an estimate of the price increase necessary to recover the cost of pollution control will determine the sustainable level of long-run demand in an industry. Awareness of this fact has focused a great deal of attention on the estimation of demand elasticities and the added costs of pollution control. The more esoteric debates raise questions regarding the application of a mark-up to added costs for profit, etc.

Yet, surprisingly, amidst this debate, little attention is devoted to the assumption that these effects, however measured, occur in the long run. In the long run, many of the argued points are simply not relevant. For example, increases in production costs due to pollution controls almost invariably rely upon engineering cost estimates of applying existing abatement technologies to existing facilities. In the long run, of course, due to technological change and, perhaps more importantly, input factor substitution (induced by the imposition of pollution controls) will serve to lower any such "static" estimate of costs.

In the case of the metal finishing industry, for example, which we consider in Chapter 7, it is difficult to believe that the long-run solution to water pollution problems will be to construct the facilities and install the equipment necessary to treat the effluent from existing metal finishing plants. Such an action would entail a cost in excess of $50 billion, or more than would be required to replace the entire capital stock of the U.S. steel industry. Needless to say, in the long run, if the metal finishing industry is to survive, new processes, products or treatment technologies will be essential. This fact is overlooked in the debates over either the accuracy of static cost estimates or the extent of any cost pass-through to consumers.

Part of our justification for focusing on short-run consequences of controls stems from the fact that, in the long run, economic resources can adjust to changed circumstances in ways which can substantially ameliorate negative effects that are apparent in a static analysis. In the short run, however, important shocks to the market may take place while the adjustment process is working itself out. In particular, one expects some plants to close, temporarily reducing industry capacity.

Just as the added costs of pollution control shift the supply curve vertically, plant closings shift the curve horizontally to the left. As plants close, each point on the supply curve shifts to the left. Since the cost of the impacted product has increased relative to other goods, a permanent contraction of the industry is to be expected. What we measure here is a larger, temporary contraction. Some existing plants will be prohibitively expensive to clean and will be closed instead. Eventually, new plants will be constructed to take their place; but in the short run, a substantial capacity "shortfall" may exist. Such a shortfall would drive prices above their long-run equilibrium value. It is this effect we have found particularly interesting.

In Figure 1-1, D_1 and D_2 represent alternative demand schedules. They are drawn so that the former passes through the horizontal portion of S (and S') and the latter through the vertical portion. Though a real demand curve might well pass through a curved section or through the vertical region of one supply curve and the horizontal region of the other, we can use D_1 and D_2 to illustrate the two pristine cases.

When D_1 is the relevant demand schedule, the impact of pollution controls is not related to the change in industry capacity caused by plant closings. The price goes from P_1 to P_1'. This price change is independent of the slope of the demand curve. The change in quantity, Q_1 to Q_1', does depend on the slope of D_1, however.

When D_2 is the relevant demand schedule, the impact is not related to the upward cost shift of the supply curve, except indirectly through the determination of plant closings. The change in production is precisely that due to plant closings, $Q_2 - Q_2'$. The price change, $P_2' - P_2$ depends on the slope of D_2.

It is a somewhat peculiar claim that the cost of pollution controls may have no effect on the market price. For example, if D_2 were perfectly elastic (horizontal), then the manufacturers would absorb all the control costs. This is possible because as we have drawn D_2, someone—possibly the manufacturers, possibly their suppliers—is collecting quasi-rents, out of which come the control costs. The question of these rents is a very important, often neglected, subject on which we will have more to say later.

It is useful to examine the problem of uncertainty about the supply schedule. At one extreme, if the industry is perfectly indifferent between new and existing equipment and new capacity can be brought on-line without delay, then the supply curve in the industry after the imposition of controls will simply have shifted in the vertical direction and the new equilibrium will occur in an essentially horizontal portion of the curve. At the other extreme, however, if pollution controls force some plants to close and it is not possible to add more capacity at any cost, then the new equilibrium might well occur on the vertical portion of the curve. In reality the supply curve is likely to be somewhere between these two extremes, but this may be very difficult to establish. However, we can bound the change by considering the extreme possibilities. The minimum impact on both price and quantity occurs when the supply curve after controls is horizontal. The maximum impact occurs when the demand curve intersects the vertical portion of the supply curve. Thus, even without observing the true supply curve, we are able to establish upper and lower bounds on the price and quantity dimensions of economic impact.

Our story is, thus, quite simple: pollution controls raise costs and also may contribute to capacity reductions. Because of these capacity constraints, it is possible, under very plausible assumptions about the characteristics of product demand, that short-run price increases may exceed the added costs of pollution abatement. In the long run, of course, these revenue increases both create an incentive and provide some of the cash necessary to build new, clean plants which can capture some of the profit. Eventually, the price could well be brought back down to where it just covers costs, including costs of abatement.

The Dynamic Implications of This Process

It is clear then that a complete analysis must take into account dynamic considerations. Such an analysis is difficult; in particular, economists do not yet have a satisfactory understanding of the investment process which ultimately determines how long it takes to reestablish an equilibrium without quasi-rents to producers—if, indeed, such an equilibrium even does exist. We limit ourselves here to a simplified picture which extends our static analysis through the time dimension. Figure 1-2 is such a representation.

In Figure 1-2, the lines D_0 to D_7 are demand schedules which show demand increasing through time, presumably in response to rising income. The solid lines S_0 through S_7 are the corresponding short-run supply curves in the absence of pollution regulation. The solid horizontal line shows the equilibrium price remaining constant through time. (This constancy is a neutral assumption that has no impact on the argument.)

The advent of pollution regulation normally causes each supply curve to be shifted upward and to the left giving S'_1 through S'_7. We deliberately draw S'_7 closer to S_7 than S'_1 is to S_1 in order to indicate that the most serious capacity impacts usually occur in the years immediately after the passage

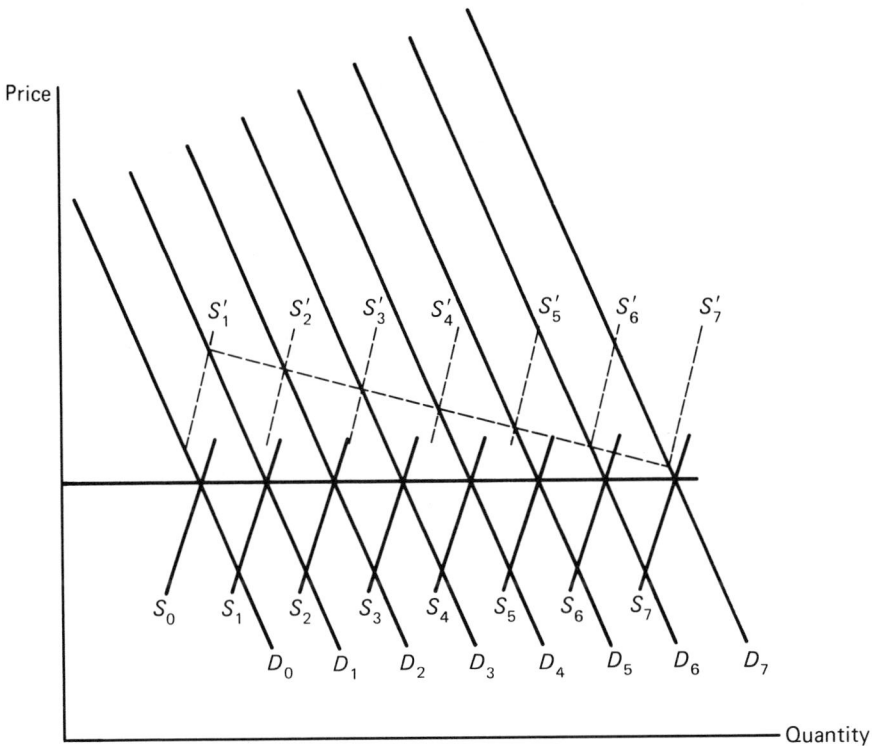

Notes: S_i, S_i': Supply curve at time i before and after the Act, respectively.
D_i: Demand curve at time i.

Figure 1-2. Transition of Equilibrium from Short-Run Supply and Demand
Curves

of the pollution control legislation. As time progresses, there will be extra investment to make up for closed plants in addition to normal expansion.

The dashed line connecting the intersections of D and S' shows the equilibrium price path over time. We call this the price "blip" out of regard for the possibly sudden increase in price above the level dictated by cost increases alone due to capacity constraints, followed by a gradual decline to the long-run equilibrium price.

It is appropriate to point out that the short-run constraint on capacity which is the source of any price "blip," can be the result of many factors besides plant closings. Most important, perhaps, is the diversion of capital, in the short run, from investment in new plant capacity to investment in abatement equipment needed to clean up existing capacity. Similarly, bottlenecks in capital goods industries, construction delays, constraints on managerial resources, etc., can all slow the long-run adjustment of an industry to the new conditions of effluent control.

In a sense, this comparative static analysis of supply and demand raises more questions than it answers. Why does the supply curve shift up by the amount it does? How many plants will close? In the presence of both abatement costs and a price "blip," who will profit and who will lose? We will employ several additional diagrams to explore these issues further.

Consequences of an Uneven Distribution of Pollution Control Costs

The starting point of almost every impact study is that pollution abatement will add X dollars to the unit cost of production in the impacted industry. We prefer instead the notion, which though obvious once stated is almost universally disregarded, that there will be a wide disparity in control costs among different manufacturers in the same industry. If there is any single theme which can be found in each of the chapters that follow, it is an attempt to measure just how abatement costs differ from one manufacturer to the next. Two important ideas are developed from this concept. The first is that the imposition of pollution controls which result in costs being unevenly distributed across producers in an industry can result in major shifts in competitive advantage within the industry. The second is that a correct industry-level analysis must employ a marginal rather than an average cost measure when translating the cost of abatement controls into price increases.

In some industries, these distributional consequences may be of far greater significance than the price and output changes observed at the aggregate industry level. Our main tool for studying these effects is what we call the "marginal control cost" schedule or the MCC.

The first step in forming this schedule is to calculate the added costs due to pollution abatement for each economic unit. In practice this has meant obtaining an estimate of the costs of cleaning up a single plant and dividing that by plant capacity. This gives us an average unit cost for each plant in an industry. It might, in principle, be better to estimate a cost function which varied with output even within a plant; but we feel that to be an unnecessary refinement. Control systems are generally characterized by a high fixed-cost component. It makes little sense to build an expensive treatment facility and then let it lie partially unused; so we expect managements to try to clean up a plant only if they can keep it fully utilized.

There are a variety of ways in which individual plant costs can be estimated. They range from direct engineering estimates on each plant to estimates for only a few "representative" plants from which some form of extrapolation must be attempted. Some of the different methods are discussed in the relevant chapters.

For our purposes, we employed engineering cost estimates, typically for representative plants; and, on the basis of important variable determinants of cost, extrapolated these costs to individual plants. The results are what we call "representative costs" for actual facilities. The true costs for any individual facility can differ from our estimates for at least three reasons: (1) inaccurate engineering costs for the representative plants; (2) unaccounted for site-specific costs; and (3) a mistaken or overly simplistic basis for the extrapolation of costs in representative facilities to individual plants.

Once a unit cost is assigned to each plant, we proceed to rank the plants in ascending order according to these costs. Thus we obtain a sequence at the head of which are usually plants which already achieve the effluent abatement levels mandated by the 1972 legislation and thus will experience no additional costs to comply with the Act. At the tail of this sequence are plants with relatively high, perhaps prohibitively high, abatement costs. Since the plants are ordered and since we know the capacity of each plant, we can determine the total quantity of industry output that can be produced at or below a given cost of abatement. Hence, for any given percentile of capacity, we can identify the "marginal" plant and its associated cost.

Figure 1-3 shows just such a marginal control cost schedule as the upward sloping, solid line. The total control cost, which is the focus of most impact studies, is the area under the MCC line. Since the MCC line is upward sloping, if we close the dirtiest X percent of capacity rather than cleaning it up, we save much more than X percent of total control costs.

Suppose we had calculated the price increase expected after the advent of abatement regulations. This is drawn on Figure 1-3 as the dashed horizontal line. B is the break-even point; hence all plants to the left of B, that is, with lower costs, will profit by the new regulations; those to the right will lose. The notion that some firms will profit because of the imposition of more stringent pollution abatement requirements is not really very surprising, for if a plant is already in compliance or almost so, it will profit as everyone else is forced to raise prices to recover the cost of new treatment facilities. The shaded area to the left of B represents the total profit to those firms which benefit from controls and, similarly, the shaded area to the right represents losses to those which lose. It is even possible that in aggregate the profits may outweigh the losses, and the industry may make a net profit.

Once an analysis such as in Figure 1-3 is complete, it is in theory

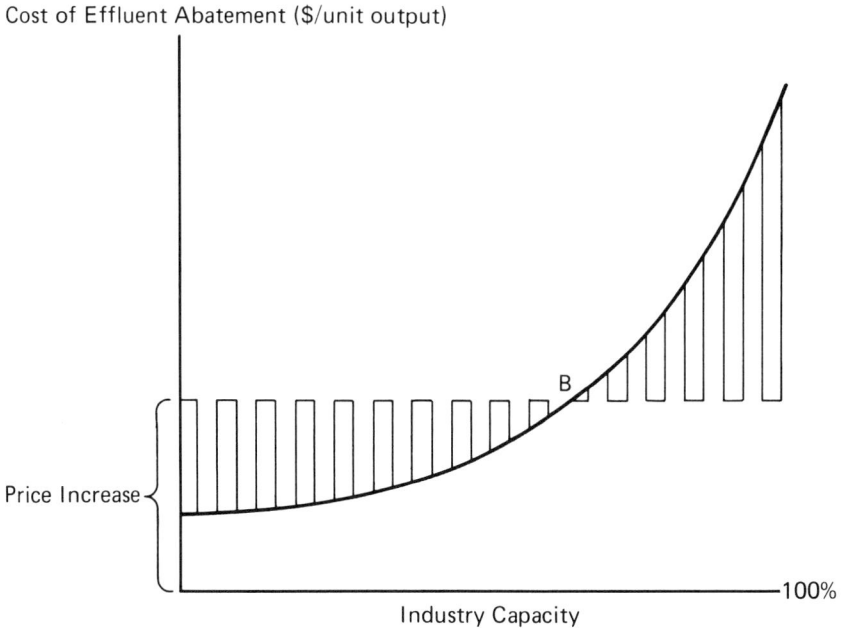

Notes: *B*: at this break-even point, the added costs of effluent control equal the increase in price sustainable in the market place.

Figure 1-3. The Marginal Control Cost (*MCC*) Curve

possible to identify a profit or loss for each production unit. These profits and losses can then be aggregated along different dimensions to examine various intra-industry consequences.

Briefly, we can mention a few of the possibilities, some of which are dealt with in specific chapters. One obvious item of interest is that of regional impact. Plants in a region often share common cost characteristics that tend toward giving them all a similar degree of vulnerability (or advantage) to any given change in production costs. This phenomenon may partly explain regional differences in political support for various elements of pollution control regulations.

Another possibility is to aggregate costs by manufacturing process. The relative efficiencies of alternative methods of production may change significantly when effluent controls are imposed. Closely related to this are changes in costs for different products. The cost analysis in Chapter 3, for example, suggests that water pollution controls greatly increase the attractiveness of unbleached or partially bleached paper products.

Another dimension that may have very important consequences, but is difficult to determine quantitatively, is the effect that an uneven abatement-cost distribution may have on industry structure. Because large companies often have newer plants and better access to capital, or because there may be economies of scale in wastewater treatment facilities, the passage of antipollution legislation may accentuate any existing tendencies toward oligopoly within an industry.

It is not only true that pollution controls may contribute to changes in industrial structure, it is also true that an industry's structure partially determines the impact of pollution controls in the first place.

For example, we have noted that the extent of any short-run capacity shortages induced by pollution controls is partially determined by the speed with which new capacity is brought on-line in an industry. Similarly, how long the low-cost firms will continue to reap the quasi-rents depends on how rapidly new, low-cost facilities replace existing high-cost facilities.

Particularly in some oligopolistic industries, quasi-rents can continue to exist for a long time. It follows, therefore, that capacity and investment decisions within an industry will be an important determinant of that industry's supply curve and, hence, of the impact of pollution controls.

Chapter 2 argues, for example, that recent circumstances in the petroleum refining industry have made timely expansion of refinery capacity unattractive; consequently, there may be periods in which capacity is limited, permitting refiners to earn quasi-rents, out of which any higher costs associated with reductions in water pollution might come.

Chapter 3 argues that the capital-intensive nature of waste treatment contributes to the development of an oligopoly in the pulp and paper industry. Such a development, in turn, has implications for capacity and investment strategies in this industry. Of course, these strategic decisions will, in turn, influence the very impact we are attempting to observe.

Costs More Carefully Defined

Up to this point we have used the term "costs" somewhat loosely. There are several different cost concepts that can be applied depending on the purpose of one's analysis.

A precise and meaningful assessment of the "impact" of a specific piece of 1972 legislation must consider both the timing of pollution abatement expenditures and the specific law these expenditures were intended to satisfy. For example, it is necessary to distinguish between those costs that are induced by the 1972 legislation and other similar but distinct costs motivated by the stringency of prior environmental legislation, both state and federal. It is also useful to identify the extent of water treatment systems already in place, because costs and impacts remaining to be incurred are reduced by a corresponding amount.

By classifying abatement expenditures according to timing and motivation, we have developed four alternative cost concepts:

1. All-inclusive costs
2. Impending costs
3. Costs due to the Act
4. Impending costs due to the Act

All-inclusive costs (1) represent the cost of taking each affected plant from a level of no wastewater treatment to a level that complies with the specific Act we are examining. Dismissing conceptual questions as to precisely what constitutes a zero treatment level, this concept of cost represents the sum of the costs to replace existing pollution control facilities plus the impending costs (2) necessary to take each plant from whatever state it is in now to compliance with the Act. It is these impending costs that are most relevant for determining the new industry equilibrium, since the impact of costs for facilities already in place has previously been absorbed by the economy, the industry, and its customers.

Costs due to the Act (3) are the costs necessary to take a plant from treatment levels consistent with previous legislation into compliance with the Act. Of course, some plants will already have in-place treatment sequences that exceed the requirements of previous legislation. These costs are, by our definition, costs due to the Act since they are in fact necessary to reach the mandated compliance levels, even though they may have been voluntarily undertaken. (They may, of course, have been undertaken in the anticipation that legislation similar to the Act would be forthcoming.) Other plants may not yet have reached standards already required by previous legislation. The costs involved in first reaching prior standards are not included in this cost concept. Thus, costs due to the Act are a measure of the expense directly attributable to the particular piece of legislation.

Impending costs due to the Act (4) are, logically, at the intersection of impending costs and costs due to the Act. They can be measured either as impending costs minus costs necessary to meet prior requirements or as costs due to the Act minus those expenditures already made in excess of prior requirements. The question, "How much more will be spent in the presence of the Act than would have been spent if the Act had not been passed?" is answered by impending costs due to the Act.

Measuring these different concepts is sometimes more easily said than done. Just understanding that a difference exists may help to clarify some discussions. One reason different parties may come up with different compliance-cost estimates is that they often measure different concepts of cost in the first place.

There are other major sources of differences in cost estimates which confound the comparison of various impact studies. First, there are differences in costs associated with differences in estimates of factor prices. Although one might think this problem rather uncomplicated, it is not in practice. As we document in our examination of the steel industry, there exist differences in state and local tax laws which, in turn, can alter the burden of pollution control costs on any given firm.

Second, differences in cost estimates can differ even when factor prices are the same if different abatement methods are elected. In several industries we attempt to demonstrate the sensitivity of option selection—and hence total cost to the industry—to price variations in several important factor inputs.

On the one hand, this analysis seems to cloud an already confused understanding of "costs"; on the other hand, we consider the mere concept of cost to be ambiguous at best and feel it is more important to identify the potential trade-offs managements face in trying to comply with the law than it is to agree on a single cost estimate.

Price and Quantity Impacts

Once we have cost estimates, two central issues remain unaddressed. We still need to know how much prices will rise in response to the control costs and how many plants wll close down rather than meet the pollution control regulations. We employ the common economic assumption that a plant will close when it cannot cover its variable costs. In this connection, the entire abatement cost is variable because the firm need not incur it; it can close instead.

Figure 1-4 presents this analysis in some detail. Suppose for each plant we were able to estimate total variable costs (VC) and total costs (TC) for production expenses just as we did previously for pollution control cost expenses alone. Ranking the plants in the same order as previously, we draw the lines marked VC and TC. The result is a set of industry marginal cost curves. In long-run equilibrium, the marginal plant must be just covering its total costs. Similarly, at any point in time, the marginal plant must be at least covering its variable costs.

We assume the industry to be in equilibrium before the imposition of the new regulations. The equilibrium point (P, Q) is at the intersection of the demand curve, D, and total cost curve TC. Notice that since we have drawn the curves sloping upwards, all the firms in the industry, except the one at

Control Cost ($/unit output)

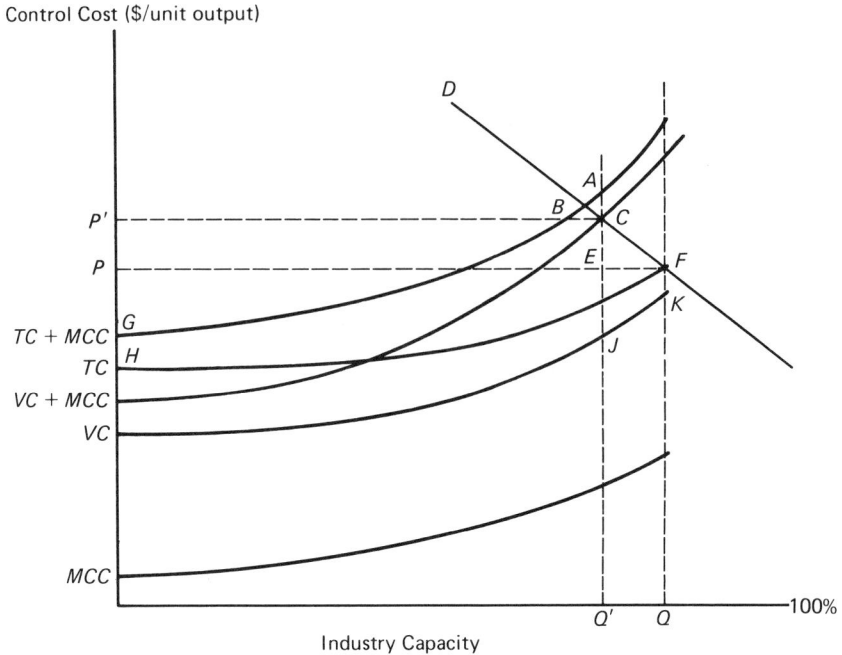

Industry Capacity

Notes: *TC* is the total production cost exclusive of effluent abatement costs.

VC is the variable production cost exclusive of effluent abatement costs.

MCC is marginal cost of effluent abatement (including both operating and maintenance and annualized capital costs).

P, Q, P', Q' are the price and quantity equilibria before and after the Act, respectively.

HFP and *GBP'* show quasi-rents before and after controls, respectively.

EFKJ and *ABC* represent losses experienced with the imposition of effluent control.

D is the demand curve.

Figure 1-4. Price and Quantity Impacts of Effluent Control Costs: The General Case

the margin, are collecting quasi-rents. We will return to the issue of these rents shortly.

We next add our *MCC* curve to both the *VC* and *TC* cost curves yielding *VC + MCC* and *TC + MCC*. The new industry equilibrium is (P', Q'). $Q' - Q$ determines the amount of capacity that will no longer be required. Since capacity was cumulated by plant, this shift to Q' determines the specific plants that will close if all costs are accurately estimated. $P' - P$ determines the price increase. Firms to the left of point B will be making a profit after the regulations while those to the right will be taking a loss, although they will be covering all variable costs.

The issue of profit must be dealt with with some care. The economist's use of "total cost," and the sense in which we use it here, includes a "fair" or "normal" return to capital. Thus the marginal plant that just covers total cost is making a profit in the accountant's books, but is making a zero "economic profit." Our use of profit is perhaps better identified with rent. Those plants to the right of B that stay in business will continue to pay dividends to their stockholders. These dividends will be less than they would have been without the Act, so we say these firms take a loss. The distinction may be important from a public policy standpoint because, while the legislature may well feel it unfair to deprive companies of profits, they may not object in the slightest to lowering "profits" attributable solely to rents.

Simply identifying the break-even point, B, in Figures 1-3 and 1-4 provides no information on the initial configuration of any quasi-rents. It may be that there are initially no rents in the system. This is quite plausible, especially as a first approximation. Figure 1-5 depicts a situation in which there were no rents prior to imposition of controls, that is, all plants have the same initial cost structure. The two shaded areas around P' are just the profits and losses associated with pollution abatement as indicated in Figure 1-3. The shaded area at the right of the diagram (E) shows the losses absorbed by those firms that choose to close down.

Returning to Figure 1-4, which allows for the existence of initial quasi-rents, we see that the area HFP represents total quasi-rents before controls. These are the rents assumed not to exist in Figure 1-5. The area GBP' represents rents after controls. The areas $EFKJ$ and ABC represent losses to closing plants (as before) and losses to plants remaining in operation, respectively. Therefore, the total net increase in profit due to controls is GBP' minus the sum of HFP, $EFKJ$, and ABC. Though we have drawn the picture so as to show a net loss, this need not be the case. However, the diagram does show that the industry will lose some part of the rents it collected prior to controls.

One may legitimately raise the question of the practicality of this somewhat involved analysis. In truth, it is unlikely that an analyst will find the information to identify the various areas. Several useful purposes may be nonetheless served. Suppose we know the MCC curve and the initial equilibrium position (P, Q); suppose we also know the VC curve in the vicinity of Q. This is sufficient to allow us to determine P' and Q', even though the various rents cannot be calculated. Further, we can make the

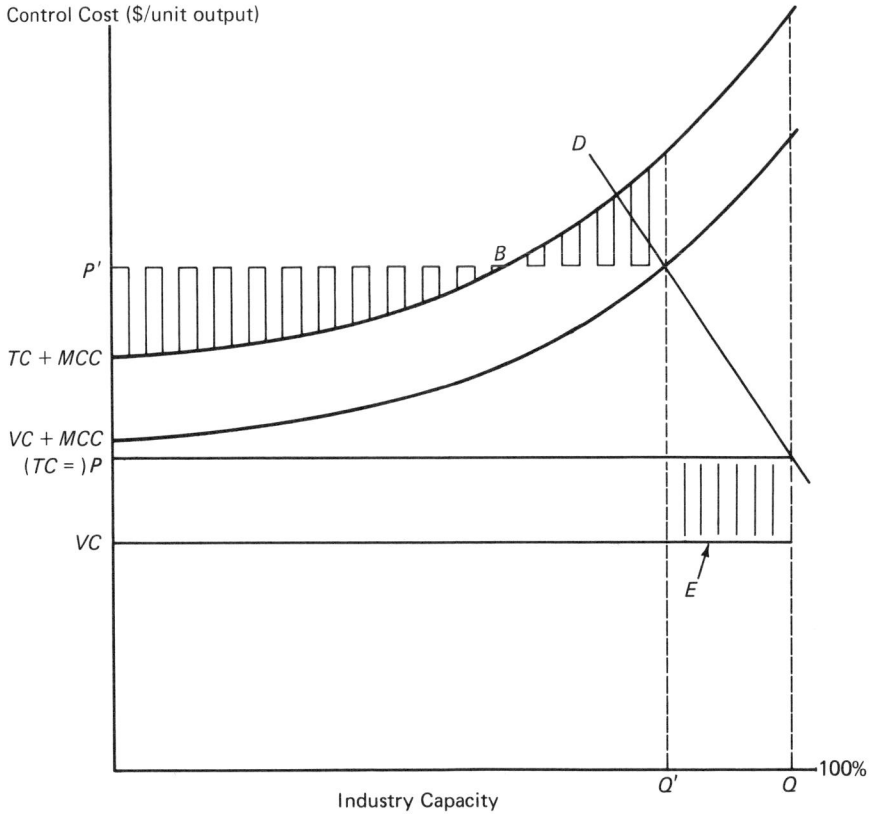

Notes: TC, MCC, VC, P, Q, P', Q', and D are defined in Figure 1-4.

           B is the break-even point as defined in Figure 1-3.

           E shows the loss absorbed by firms that close in response to the Act.

Figure 1-5. Price and Quantity Impacts of Effluent Control Costs, Assuming No Quasi-Rents Exist Before Control

assumption that there were initially no quasi-rents, as in Figure 1-5, and calculate the indicated profits and losses.[c] For most purposes, knowing the new equilibrium price and quantity and having an approximation to profit or loss (and perhaps knowing the direction of bias) will be sufficient. Further, even though as analysts we may not be able to identify the entire total and variable cost curves, individual managements probably know at

---

[c] If the rank ordering among plants of abatement costs is the same as the rank ordering of production costs among plants, we then have an upper bound on net profit (or a lower bound on net losses) to the industry attributable to pollution controls.

least their own positions within the industry. They are then in a position to make use of our results as a valuable planning tool.

This completes the description of our basic model. We are able, at least in theory, to determine profits and losses due to the imposition of pollution regulations. Several caveats are in order, however. For instance, these profits do not necessarily accrue to the manufacturers themselves. The rents in the system may well be attributable to some factor supplier. Traditionally, rents are associated with a resource in scarce supply, not always a manufacturing process. It is quite possible that what appears in Figure 1-5 to be a loss (or gain) to the industry may actually be passed back to a supplier. Gains might be translated into inflated capital goods prices, for example.

This rent pass-through is very difficult to estimate in any real application because, in addition to everything else, we now must know the supply curve for input factors. The effect may be very real, however, and very important. Several of the industries reported on here are primary raw material processors. It is quite likely that some of the losses associated with pollution control will be borne by OPEC, by the owners of forests, and by mine owners.

One might still ask why the *MCC* curve is upward sloping. If it is simply because some facilities are more easily brought into compliance than others, then the analysis is as stated. It is possible that shortages of pollution control equipment may occur. In this case some of the captured profits are passed on to the producers of such equipment. While another diagram could no doubt be produced to show this, we will leave it as an exercise for the reader.

As economists customarily do, we have assumed perfect competition throughout this chapter. This simplifies our diagrams, but may not accurately describe some of the industries we have studied. An analysis of the oligopoly case is clearly called for. While economists do not have a really satisfactory framework for such an analysis, we can study the limiting case of a monopoly. Figure 1-6 presents the analysis. (We abstract from the difference between variable and fixed cost.) The marginal cost, marginal revenue, and demand curves are labeled *MC*, *MR*, and *D*, respectively. After the passage of pollution regulations, the cost curve is *MC* + *MCC*. In general, the price increase $P' - P$ may be either more or less than the *MCC* at the new output level Q'. However, the monopolist suffers an unambiguous loss. The area *ABCD* is the increase in revenue due to a reduction in sales. The loss is necessarily greater than the gain. (If not, the monopolist could have profitably raised prices before the imposition of the pollution regulations.) The cross-hatched area *HJKL* represents the cost of abatement; this cost is borne solely by the monopolist. We could repeat the previous discussion as to whether the shape of the marginal cost curve is

Control Cost ($/unit output)

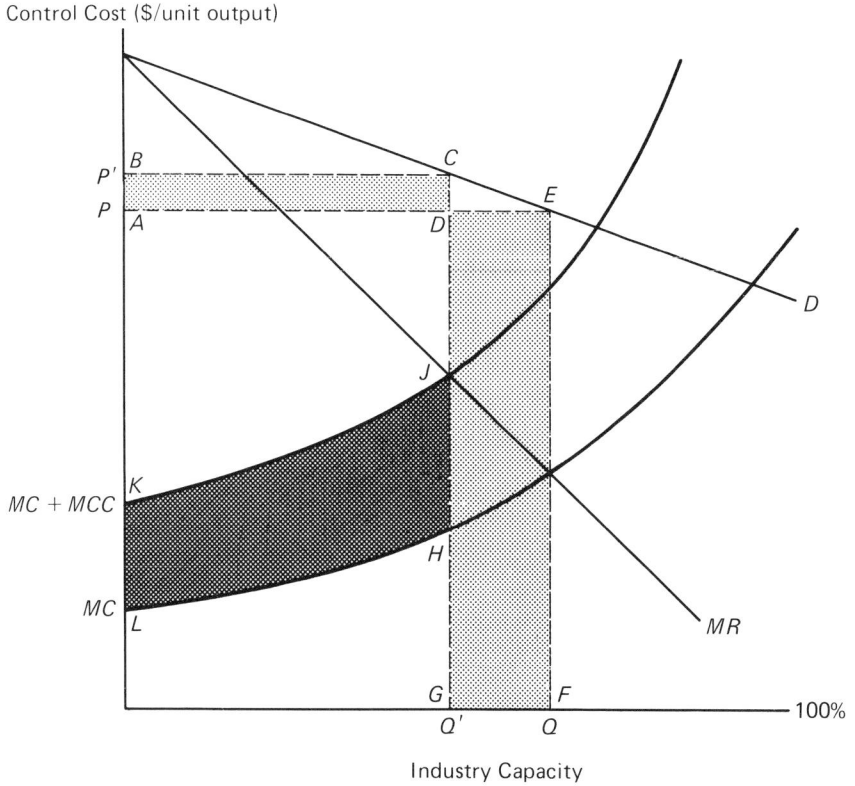

Industry Capacity

Notes: P, P', Q, Q', MCC, D are defined in Figure 1-4.

           MC, MR are the marginal costs and marginal revenue, respectively, experienced by the monopolist.

           $ABCD$ is the increase in revenue associated with the higher prices after the Act.

           $DEFG$ is the loss in revenue associated with the reduction in sales after the Act.

           $HJKL$ is the cost of abatement to the monopolist.

Figure 1-6. Price and Quantity Impacts of Effluent Control Cost: The Monopolistic Case

determined by a rent to a supplier. We could also point out that both in this case and the previous one, the change in rent would be different if the factor supplier were a monopolist.

Summary

Our description of the impact of environmental controls on industry is based on simple principles of microeconomics. Pollution controls increase production costs and may contribute to capacity reductions. The resulting shift in an industry's supply curve, when coupled with a sloping industry demand curve, determines both the price and quantity impacts of pollution abatement on the industry.

As a rather simple exercise in comparative statics indicated, it is possible, under very plausible assumptions about the nature of supply and demand, that short-run price increases may exceed the added costs of pollution abatement.

We also noted that the burden of pollution abatement costs is typically borne unevenly by competitors in an industry. Thus, pollution control costs constitute a source of quasi-rent to some industry producers. We also noted, however, that the existence of quasi-rents, due to other factors prior to the imposition of effluent controls, might well compel some firms to absorb at least some of the added costs of environmental improvement, thus ameliorating the price and quantity impacts of the legislation.

We also observed that in the long run, revenue increases associated with any quasi-rents created by the imposition of an uneven cost burden on an industry both create an incentive and provide some of the cash to build new, clean production facilities. Eventually, we concluded, the price could well be brought back down to where it just covers costs, including costs of abatement.

This dynamic adjustment to any new, long-run equilibrium depends on many poorly understood industry phenomena, including the strength of competition and the investment pattern of firms in the industry. Pollution controls, by modifying an industry's cost structure, can change the very mechanisms that ultimately determine the impact of the pollution controls on the industry.

2 The Petroleum Refining Industry

James L. Smith

Several favorable circumstances establish the petroleum refining industry as a good laboratory for the application of the economic impact methodology sketched in Chapter 1. Estimates of water pollution abatement costs for this industry are among the most complete and detailed available, permitting the identification of impacts as they are likely to be felt on an individual plant level. In addition, the estimates have been calculated in a way that isolates the roles played by changing product mix and process characteristics in determining the level of abatement costs. Consequently, it is possible to record the effects of the fluctuations in these variables on alternate cost estimates.

The level of detail in the impact study also reflects the fact that many aspects of the domestic petroleum refining industry are open to public scrutiny. A comprehensive census of all U.S. refining operations is published annually in *The Oil and Gas Journal*. The census reveals the size, process characteristics, product mix, location, and ownership of every operating refinery. We have made full use of this material, supplemented by additional information describing the cost and distribution of wastewater treatment facilities within the industry. Consequently, the correspondence between estimated plant level costs and the real costs actually to be experienced by each plant is believed to be stronger in the refining industry than in any other.

Application of our research methodology to the refining industry is greatly simplified by the essentially competitive nature of the industry. However, the reader should not interpret this to mean that we have treated all segments of the petroleum industry as behaving according to a competitive model. We have not. There are obviously monopoly elements in the area of crude oil production, and we have explored these elements as they relate to the problem of pollution control. In addition, the presence of strong brand names and product loyalties in the retail marketing segment of the industry suggests a model of monopolistic competition. However, our focus is on the refining segment, where the problem of water pollution occurs.

The structure of the refining segment is markedly different from either the production or retailing segments. Crude oil is refined at 250 plants throughout the U.S., with a combined rated capacity of over 15 million barrels per day (mbpd).[1] The size of refineries varies substantially from 250 bpd to over 400,000 bpd. Ownership of refineries, while dominated by a small number of large integrated firms, is nevertheless fragmented among nearly 200 independent companies.

Wholesale market areas of individual refineries overlap considerably. Almost all refineries compete to some degree in a nation-wide petroleum products market via the integrated circuit of pipeline, barge, and tanker transport. The products of refineries are undifferentiated at the wholesale level. A particular fuel meeting a given set of technical specifications is a neutral, unbranded commodity. Only at the retail level of distribution do brand names succeed in differentiating petroleum products. Because refinery products are identical in the wholesale market, price can be the only distinction, and consequently, it is the primary basis for competition.

Strong competition among domestic refiners has been observed often in the past. Because of major economies of scale, expansion in refining capacity has typically come in quantum increases during periods of favorable economic conditions, rather than in sustained incremental growth over time. As a result, the industry has tended to oscillate between periods of under- and overcapacity. Periods of surplus productive potential (more products can be produced at a given price than can be sold) have often been marked by sales discounts and strenuous price competition, rather than by curtailing industry output. It rarely benefits an individual refiner to reduce production, even when prices are low. As long as price covers variable costs—which are extremely low in refining—it pays the refiner to operate near full capacity. It would require collusive agreements to achieve a coordinated production cutback during times of a weak market, and this has not been observed.

Thus, we assume that the structure of the refinery segment of the petroleum industry is reasonably competitive, that is, that strong forces exist which tend to make the price charged for refinery processing gravitate to the marginal processing cost. However, we realize that market situations occasionally arise which fall beyond the narrow bounds of this hypothesis. For example, market disturbances emanating from member nations of OPEC, price disturbances caused by federal regulation of the petroleum industry, and changes in the product mix and process characteristics of refineries are intrusions which periodically alter the price/cost relationship in the industry. We will introduce these factors whenever it seems appropriate.

Abatement Costs Facing the Industry

Our analysis starts with estimates of water pollution abatement costs to be incurred by individual refineries. As explained in Chapter 1, we have specified four alternate concepts of abatement cost, depending upon the timing and motivation of the expenditures: all-inclusive costs, impending costs, costs due to the 1972 amendments to the Federal Water Pollution Control Act, and impending costs due to the Act.

Our practice is to estimate costs individually for each refinery, and then aggregate these costs to yield industry totals. The estimate of abatement costs at a given refinery reflects a variety of control measures induced by the legislation, as well as many operating characteristics unique to the specific refinery. We examine a wide range of abatement measures encompassing sour-water stripping, storm- and ballast-water handling, filtration, biological treatment, conversion to recycle water cooling systems, spent caustic neutralization, activated carbon absorption, and freeze crystallization to segregate remaining solids from the wastewater flow.

Costs associated with each abatement measure are broken down into capital, operating, land, and energy components.[2] Capital costs are annualized by assuming a 15 percent rate of return on investment and a 20-year physical depreciation life. Table 2-1 reports cost estimates by component aggregated for the entire industry, according to each of the four cost concepts.

From Table 2-1 we can see that domestic refiners face impending annual abatement costs due to the Act that will amount to $123 million by 1977, $332 million by 1983, and $419 million by 1985. These totals may be translated to average unit costs of 2.4¢ per barrel refined in 1977, 6.3¢ per barrel in 1983, and 8.0¢ per barrel in 1985.[3]

As our discussion of the concept of the *MCC* curve in Chapter 1 indicated, we are interested not only in the average level of abatement costs (calculated over all refineries), but also in the differences in abatement costs experienced among the individual refineries.

Although the average 1977 unit cost is 2.4¢ per barrel, some refineries will actually incur no additional abatement costs—that is, their facilities already comply with 1977 standards.[a] The unit cost of abatement rises continuously but slowly above zero as we move to the right along the marginal control cost curve. No dramatic departure from the average cost level occurs until we approach the last one percent of refinery capacity, that which is subject to the highest unit costs of abatement. This last one percent of capacity consists of many small refineries which typically produce specialty products in isolated geographic locations. The *MCC* curve for the petroleum refining industry is plotted in Figure 2-1.

We reiterate the importance of estimating the differential incidence of abatement costs, as in Figure 2-1, and distinguishing marginal abatement costs from the average. This approach provides the basis for a discussion of

[a] Unless otherwise specified, we deal in the remainder of this chapter with impending abatement costs due to the Act, since the impact of expenses previously incurred has already been absorbed by the industry and its customers.

Table 2-1
Water Pollution Abatement Costs: Petroleum Refining Industry
(in millions of 1973 dollars)

Cost concept	1977 Abatement Level (BPT)			1983 Abatement Level (BAT)[a]			1985 Abatement Level (EDOP)		
	Capital Cost	O & M Cost[b]	Annual Cost[c]	Capital Cost	O & M Cost[b]	Annual Cost[c]	Capital Cost	O & M Cost[b]	Annual Cost[c]
All-inclusive costs	1278	222	426	1805	347	636	1999	403	723
Impending costs	540	161	247	1067	286	457	1261	342	544
Costs due to the Act	969	115	270	1496	239	478	1689	296	566
Impending costs due to the Act	386	61	123	913	186	332	1107	242	419

[a]All costs are reported on a cumulative basis; hence, the 1983 costs include the costs to achieve the 1977 abatement level.

[b]Annual operating and maintenance costs.

[c]O & M plus annualized capital costs assuming a discount rate of 15% and a 20-year depreciation life.

Source: James L. Smith and Robert A. Leone, report to the National Commission on Water Quality, June 1, 1975, Section II.

Unit Cost ($/Bbl)

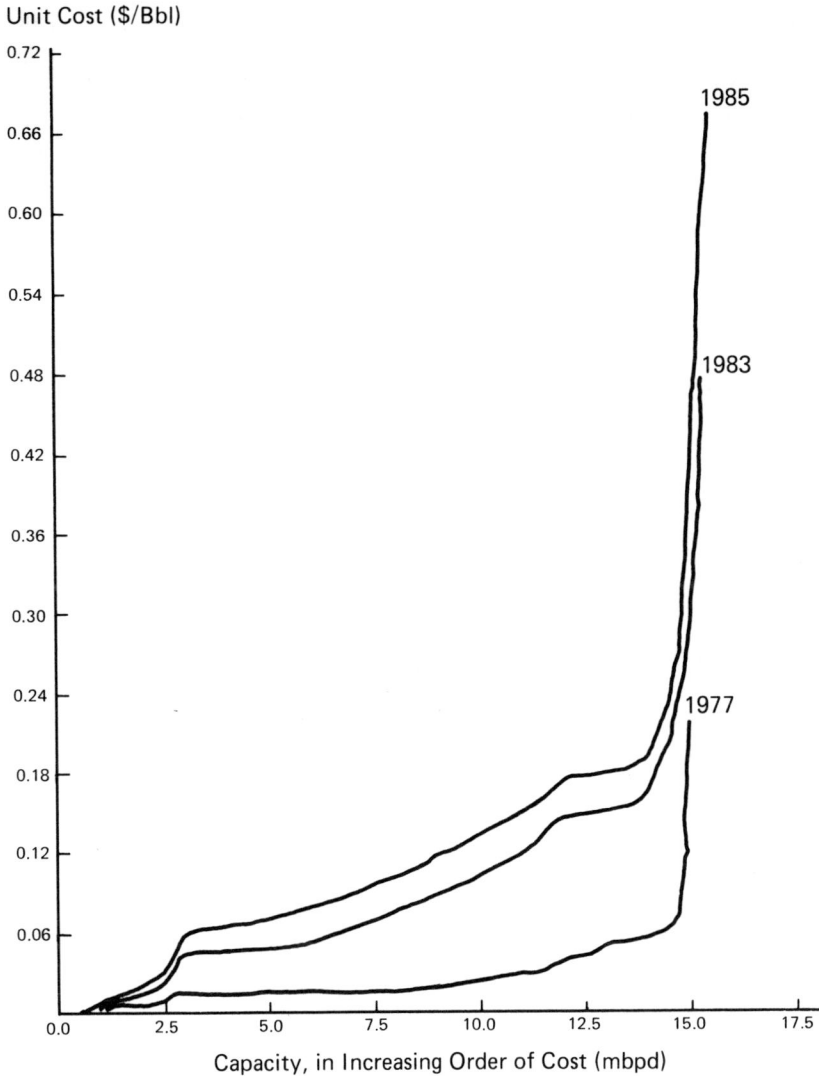

Capacity, in Increasing Order of Cost (mbpd)

Notes: Upper curve: 1985 (EDOP) Abatement levels
                      Middle curve: 1983 (BAT) Abatement levels
                      Lower curve: 1977 (BPT) Abatement levels

Figure 2-1. Marginal Control Cost (*MCC*) Curves: Petroleum Refining
Industry

the gains and losses from pollution controls within the industry, and also
enables the researcher to examine the effect of marginal abatement costs on

industry prices and profit levels. For example, we will later see that the economic consequences of pollution abatement for refinery capacity located in the high-cost tail of the cost curve may be moderated by the fact that this group consists mainly of plants producing particular product types and operating in limited areas.

Application of the Impact Methodology

Having estimated the cost of pollution control, it would be tempting simply to assume that all costs are passed through to the consumers of petroleum products, and then to calculate the negative reaction on consumption and production implied by an estimate of the price elasticity of demand. As the discussion in Chapter 1 showed, this is an overly simplified view of the consequences of pollution abatement.

The degree to which abatement costs are passed through to customers depends critically upon the conditions prevailing in the market for petroleum products at the time control measures are implemented. Even in a competitive industry there are times when the close tie between the cost of production and price is severed. Typically, this occurs during periods of capacity shortage or surplus. Although a competitive market tends to a long-run equilibrium where capacity adjusts closely to demand, there is no rule that prevents short-run fluctuations during the process of adjustment.

A discussion of the capacity problem is particularly relevant to the refinery industry because many periods of capacity disequilibrium have been observed historically, mainly on the side of surplus. Although excess capacity is a short-run problem, the duration of the short run is determined by the depreciation period of a long-lived capital investment; the surplus refinery will remain in operation until price has fallen below the level of variable costs, which are themselves very low during the refinery's useful life. The usual result is price competition in which the price collected for performing the refinery service falls well below the level that recovers the full cost of capital employed by the firm, that is, price falls below the level of competitive equilibrium.

If price falls below total cost, does this mean that pollution abatement costs will not be passed on to customers? Is the force of competition so strong in the face of excess capacity as to prevent any cost pass-throughs? The answer is clearly no, for excess refinery capacity does not prevent pass-throughs. As Chapter 1 noted, during the implementation period for effluent limitations, pollution abatement expenditures (even for capital equipment) are considered by management to be variable costs.[b] Since

---

[b] This is so, because the abatement expenditure can be avoided, just like any variable cost, by ceasing production.

managements will not incur variable expenses unless they anticipate that prices will return the higher cost, and because the cost is applied over the entire industry (albeit unevenly), there is no need for any manager to absorb the (variable) abatement cost internally. The force of competition will not erode price below or prevent it from increasing to the level which returns the abatement cost to the firm.[c]

The converse case, a capacity shortage, raises doubts about the industry's ability to pass pollution abatement costs on to consumers. A capacity shortage implies that existing refineries are unable to satiate demand at the competitive price. Prices, therefore, rise to choke off unsatisfied demand. By this process, a quasi-rent accrues to the owners of existing plants.[d] Abatement costs, if imposed at a time when quasi-rents are prevalent among refiners, merely displace a portion of the rents, thus, assuming the role of the rent in maintaining price at a level that staves off demand.

In a competitive industry offering free entry, quasi-rents are expected to attract additional capital and alleviate the shortage rather quickly. However, several characteristics of the economic environment of the refining industry threaten to inhibit timely capacity expansion.

A refinery is a long-lived, rather inflexible, capital investment. Prospects of continued demand for refinery products and a stable supply of refinery feedstock must be secured before management will commit capital to such an enterprise. The typical response to the inherent risk has been vertical integration. Firms integrate backward into crude production to secure a guaranteed supply of feedstock, and to gain preference during tight market periods. Firms also integrate forward into retail distribution to insure that established marketing channels are available to provide a secure outlet for the refinery's products.

Industry managers familiar with the integrated structure of the petroleum industry recognize the dangers of excess refining capacity. It is well known that the relatively low variable costs of operating a refinery motivate management to operate near full capacity even when prices are falling below total cost. This action floods the distribution channels during periods of a weak market. Although it is in the financial interest of each individual refiner to react this way, it eventually causes surplus products to be dumped on the spot market to be sold to independent retail distributors. The consequence of excess refining capacity, then, is to cause a high

---

[c] Of course, after abatement capital is once put into place, it is no longer regarded as variable. The manager would, then, be theoretically willing to bid the price down still lower. However, this would reflect a singular lack of foresight on the part of management; if the abatement capital were not expected to return its cost, it should not have been employed.

[d] These quasi-rents may be viewed either as "unearned profit" or as an incentive to expand. The controversy which rages between these opposing views, although interesting, is not germane to our analysis. The relevant aspect is that the quasi-rent drives a wedge between cost and price.

volume of products to move very cheaply outside established distribution channels. This invites extreme price competition between the independents and integrated distributors, who see their surplus refinery product coming back to haunt them. As the price competition is carried to its extremes, there is a danger that brand loyalties will be substantially eroded.

The fact that refined products do move freely at a price discount supports our hypothesis that the refining industry is basically competitive. However, this form of competition clearly leads management to resist expanding capacity until it is very certain that most of the production can be marketed through integrated distribution channels. A bias against early expansion of capacity can result.

In addition to commercial risk, other equally important uncertainties delay refinery construction until strong market signals (quasi-rents) indicate expansion is overdue. Debate over the security and price of expanded supplies of Mideast feedstocks is one of the most visible discouragements to capacity expansion. Debate over federal import tariff and quota schemes exerts an analogous inhibiting force. Management will not commit money to a refinery that will later be made superfluous by a cutoff of crude oil supply.

The strength of future demand for petroleum products in the wake of 1973-1974 price increases is also widely debated. We know that demand elasticities are greater in the long run than the short run; consumer responses to price changes are delayed by the force of habit, by the requisite time to discover means of conservation and substitution, and by the time consumers require to depreciate existing oil-fired capital equipment. The economic viability of alternate energy resources (particularly coal and nuclear power) is another unknown which clouds forecasts of future petroleum demand.

Until these uncertainties are resolved or hedged in some way, management can be expected to implement refinery expansion plans hesitantly, being spurred on mainly by the promise of lucrative quasi-rents.

Many of the uncertainties that currently inhibit refinery expansion can be traced to the erratic behavior of Middle East oil-producing countries. Thus, it is interesting to note that the dictated rise in price of Middle East crude oil dampens the demand for U.S. refinery capacity at the same time it reduces the supply. The demand for refinery processing services stems from consumer demand for refined petroleum products. Any arbitrary increase in the price of one component of refined products (e.g., crude oil feedstock) reduces the derived demand for the other component of those products (e.g., refinery processing).

This relationship and the role which OPEC plays in creating it, are indicated in Figure 2-2. Total demand for refined products (at the refinery gate) is represented by the line segment labeled D. The price of crude oil

Price of Petroleum Products

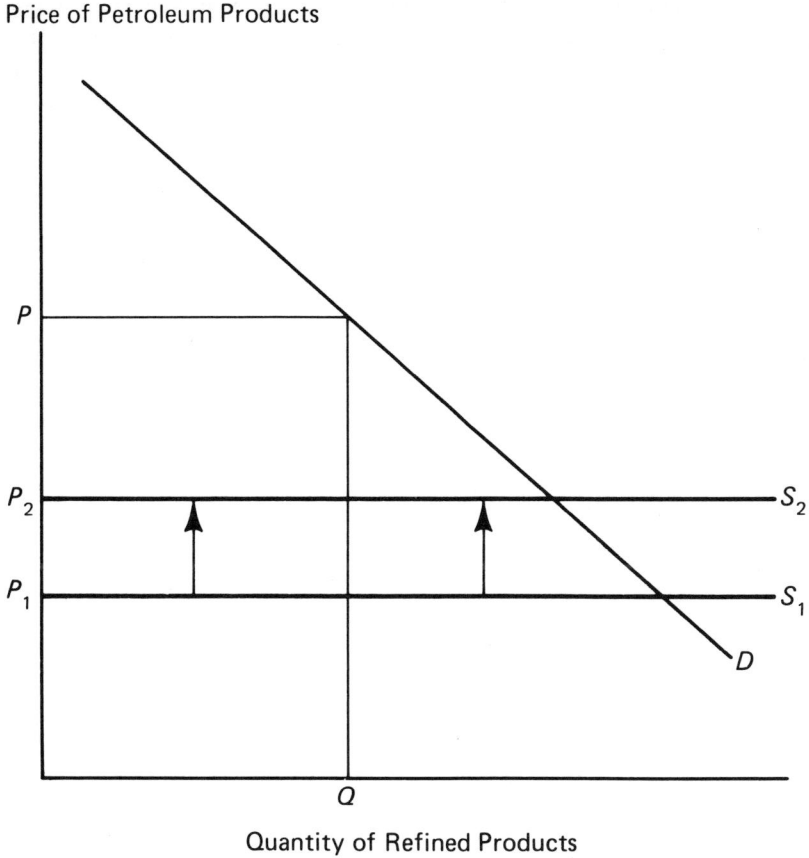

Quantity of Refined Products

Notes: D is the demand curve for refined petroleum products.

S_1 is the initial supply curve for crude oil which, with D, yields an initial equilibrium price of final output equal P.

S_2 is the supply curve for crude oil after the crude oil suppliers (e.g., OPEC) arbitrarily raise the price of crude oil to P_2.

P is the typical price of refined petroleum products, hence, $P - P_1$ (or P_2) is the refiner's revenue net of crude oil costs.

Q is the level of refined product demand at a price P.

Figure 2-2. Effect of Crude Oil Prices on Demand for Refining Capacity

feedstock is initially assumed to be P_1. Of course, P_1 represents a cost subsumed in the selling price of refined products. The demand curve specifies that at a typical price of refined products, P, consumers purchase the quantity Q. Crude oil suppliers would receive a price P_1 for their crude, and the remaining portion of price ($P - P_1$) reverts to the refiner.

Now, imagine that the crude oil supplier (an OPEC member) arbitrarily raises the price of crude oil from P_1 to P_2. After this change, the portion of product price accruing to the refiner falls by an amount $(P_2 - P_1)$; this portion of the revenue is essentially transferred from the refiner to the crude supplier.

We cannot say that the refiner would be willing to supply his services at the new lower price, but we do know that if he is to sustain the former level of demand for his service, he must charge a correspondingly lower price. This is tantamount to saying that the demand for refinery services has fallen by the amount of the dictated crude oil price increase. As the OPEC member exerts additional upward pressure on the crude oil price, the derived demand for refinery capacity is reduced further; petroleum products are increasingly priced out of the market.

Summarizing, we see that the recent events and uncertainties surrounding the petroleum industry have set in motion two countervailing forces: a reduction in new capacity formation, and a reduction in the demand for new refinery capacity. Whether these forces will result in net excess capacity or not depends upon the relative strength of the countervailing forces.

Our best estimate is that the reduction in demand will more than compensate for lagging expansion plans by 1977, when the effluent limitations take force. Although precise forecasts of future demand depend upon the price of petroleum obtaining in the interim, the Federal Energy Administration estimates that at current crude prices, domestic demand for petroleum products would fall in the 17-18 mbpd range in 1977. A significant portion of this demand would have to be imported from foreign sources, partly in the form of crude oil, partly in the form of products. If the crude/product import split remains evenly divided, as it has been recently, then the derived demand for domestic refinery capacity is estimated to range from 11.7 to 12.1 mbpd, depending, again, on the price of crude oil.[4]

Actual refinery production runs in 1974 amounted to 12.8 mbpd, significantly higher than the forecasts of future demand. Relatively low levels of demand reflect anticipated consumer substitution away from recently more expensive oil. The capacity requirement also reflects an assumed U.S. access to imports of refined products. If either of these conditions fails to hold, demand for U.S. refineries would be higher. For example, if the demand elasticities used by the Federal Energy Administration were too high, consumers would not be so successful in their attempts to avoid consumption of petroleum. Or, if import controls restrict our ability to import refinery services (as embodied in refined products), demand would again be higher.

Capacity, however, will not remain fixed at the 1974 level. The actual amount of new refinery capacity coming on-stream by 1977 is already known with some accuracy. Much of this increment has already been designed and financed. The U.S. Bureau of Mines reported 1.3 mbpd of capacity under construction as of January 1, 1974.[5] Assuming a three-year lead time to completion and a 90 percent utilization rate, the industry would be able to provide an incremental 1.2 mbpd of refining capacity by 1977.

This increment to capacity provides a sizable margin of error for our earlier calculation of excess capacity. We may reasonably conclude that even if the estimates of future demand are erroneously low (perhaps by as much as 2.0 mbpd), it nevertheless appears that sufficient capacity will be made available to avoid any significant capacity shortage in the industry. Consequently, the prospect of quasi-rents accruing to marginal refiners between now and 1977 appears dim.

Having discussed in some detail the role played by changing market conditions in determining the impact of abatement costs on price, production, and profits in the refining industry, we are left with the rather uncomplicated conclusion that the advent of pollution abatement costs is most likely to be reflected in a full cost pass-through to the industry's customers. There is no reason to expect either quasi-rents which will absorb the abatement costs or abnormal competition which prevents the cost from being passed on.

Although this conclusion is not remarkably different from the a priori notion one might have about industry response to abatement costs, the discussion serves the useful purpose of validating intuition. Moreover, our discussion of market influences indicates the potential sensitivity of pollution abatement impacts to several quite volatile characteristics of the petroleum market.

Impact of a Cost Pass-Through

Recalling our earlier estimates of the unit costs of abatement, we anticipate the size of the cost pass-through to be determined by the marginal control cost curves (see Figure 2-1). The relevant feature of those cost schedules is that most (93 percent) refinery capacity can be treated at the 1977 abatement level for less than 5¢ per barrel; at the 1983 abatement level, for less than 11¢ per barrel; and at the 1985 abatement level, for less than 14¢ per barrel. Accordingly, typical price increases are expected to be on the order of these respective magnitudes.

Price increases will be spread over a variety of refinery products, varying from asphalt to jet fuel. Each product satisfies a distinct demand with a distinct demand elasticity. To precisely predict the effect of a price rise on demand, one must know the distribution of the increase among respective products, and the elasticity in each product market.

A number of estimates of price elasticities appear in the literature. The most comprehensive estimates, published by the Federal Energy Administration,[6] indicate that the elasticity for various petroleum fuels ranges from −0.3 to −1.8. These are long-run elasticities meant to allow for a maximum amount of substitution away from petroleum products.

To simplify the analysis, we assume that each refinery product shares the price increase in proportion to its current price. Furthermore, we assume the true price elasticity for each product equals −2.0. This is admittedly a pessimistic assumption as far as the petroleum industry is concerned, but it has the virtue of identifying the maximum impact that can reasonably be expected in the industry, in terms of production curtailments and plant closures.

In the event of the hypothetical cost pass-throughs (5¢, 11¢, 14¢), a demand elasticity of −2.0 would imply a reduction in consumption of 0.10 mbpd by 1977, 0.21 mbpd by 1983, and 0.28 mbpd by 1985.[e] Such reductions, which constitute approximately 1 percent of capacity, would be barely perceptible when distributed over the entire industry. In other words, the strength of consumer demand for petroleum products appears sufficient to absorb the costs of pollution abatement, even under the most unfavorable assumptions. We conclude that, under the competitive model, only an extremely small percentage of domestic capacity could be priced out of the market by water pollution abatement costs.

These results are hardly surprising. While we entertain abatement costs varying from 5¢ to 14¢ per barrel, the average barrel of refined product is selling for more than $18.00. Abatement costs constitute far less than 1 percent of the current price of the final product—an amount that should have only a trivial effect on demand.

Industry representatives may be concerned that, in spite of harmlessly low unit costs of abatement, expenditures for treatment facilities will divert productive capital from expansion of basic refinery capacity. According to the argument, there is some danger that implementation of abatement measures will cause a "capacity crisis" all its own. We see this danger as being minimal.

Prior to 1977, the negative force of abatement expenditures on new capacity formation will be small since the increment to capacity is reported to be already under construction. Moreover, impending 1977 capital requirements for abatement control in the industry are on the order of $390 million (Table 2-1), which translates into approximately 0.195 mbpd of new refinery capacity.[f] Even if this 0.195 mbpd were entirely foregone, we

---

[e] Assuming a base level of consumption of 17.3 mbpd in each case. This figure can be varied without materially affecting our conclusions.

[f] This conversion is based upon the estimate that new refinery capacity entails capital expenditures of $2 billion per million barrels of daily capacity. See E.K. Grigsby, E.W. Mills, and D.C. Collins, "Refiners Facing Future Need for $5.3 Billion/Year Investments," *Oil and Gas Journal,* May 7, 1973.

believe that the U.S. could import the difference in the form of refined products without putting a perceptible strain on world refining capacity. But, again, we emphasize that we do not anticipate the need to forego any of the 0.195 mbpd of capacity.

The cumulative capital invested in water pollution abatement equipment through 1983 and 1985 is estimated at $0.9 billion and $1.1 billion, respectively (Table 2-1). This monetary investment is the equivalent of an investment in 0.450 mbpd of capacity by 1983, and 0.550 mbpd by 1985. These quantities represent approximately 10 percent of the capacity expansion that is apparently required to accommodate projected petroleum consumption over the next ten years.[7]

It is not clear that undertaking pollution abatement investments actually displaces the 10 percent of capacity that otherwise would have been forthcoming. The degree of displacement depends upon society's propensity to save and on the profitability of proposed capacity expansion. In the extreme case, however, if we assume that all 10 percent is foregone, it still appears likely that the shortfall of 0.55 mbpd could be made up through the importation of refined products.

Plant-Level Distribution of Impact

Although the aggregate industry-level impact of pollution abatement is relatively small, this does not mean that there will not be individual refineries affected strongly by the legislation. The sharp "tails" on the marginal control cost curve (Figure 2-1) reflect an unequal distribution of abatement costs among plants that will put some at a competitive disadvantage.

A primary conclusion of the preceding industry-level analysis was that the strength of demand for refined products is sufficient to absorb almost all of the costs of water pollution abatement. Our assessment of the most extreme industry impact that might be expected suggests that no more than 1 percent of domestic refinery capacity would be priced out of the market. However, rather than limiting the analysis of plant-level impacts to the last 1 percent of capacity (as suggested by the strength of industry-level demand), we treat the last 2 percent of capacity in order to deal more comprehensively with plants in the high-cost tail of the marginal control cost curve.

The 2 percent of capacity subject to the highest unit abatement costs consists of 49 refineries, typically small in size. These refineries are scattered widely over the U.S.: 12 in the East, 9 in the Midwest, 18 in the South, and 10 in the West. It would be tempting to conclude simply that our group of 49 potentially impacted refineries is representative of those that will lose market share, go out of business, or be forced to accept less than a fair return to their capital. As our discussion in Chapter 1 indicated, this simplistic conclusion is very likely to overstate the true short-run impact of pollution control; for the market position of a potentially impacted refinery is protected in the short run by lags in the enforcement of competitive pressure and by rigidities occurring on both the supply and demand sides. For instance, although customers served by a high-cost abater may have the option of switching to alternative fuel sources in the longer run, they may nevertheless be forced by their own physical capital to continue using a specific fuel or fuel type in the short run.

Apart from the lags in competitive pressure and readjustment that simply forestall the inevitable, there are other factors that appear to protect the competitive position of potentially impacted refineries. At least four sources of potential economic rents may sustain the high-cost refiners: (1) servicing of a small, geographically isolated market; (2) ownership of a favorable local crude oil source; (3) production of specialty products (e.g., asphalt) which face only local competition because of prohibitive transport costs; and (4) a general capacity shortage among refineries in a given market area. When any of these factors bolster profit margins, additional costs of pollution abatement can be absorbed without depressing the profit level below the competitive rate, and the potentially impacted refinery may adjust internally without raising prices and endangering its competitive position.

Geographic Impact of Water Pollution Controls

With these qualifications in mind, we examine our list of potentially impacted refineries by focusing on the geographical proximity of competitors and the similarity or dissimilarity of competing product lines. This discussion proceeds region by region to illustrate the importance of unique factors in determining the likely economic impact of water pollution controls on specific refineries.

The Atlantic Coast

There are twelve potentially impacted refineries in this region, nine of which produce specialty lubricating-oil products almost exclusively. These nine, concentrated mainly in western Pennsylvania and West Virginia, clearly compete with each other for markets but not necessarily with "nonlube" refineries in the area. Because no other refineries can provide comparable lubricating products at lower abatement costs, there is no reason to believe that any of the nine potentially impacted "lube" refineries will be driven out of business. Rather, since abatement costs within this narrow segment of the industry are relatively uniform, the price of lubricating oils will probably rise in greater proportion than the prices of other petroleum products. Impact in this case is likely to result in product price increases rather than a change in the composition of the industry.

Besides the nine "lube" refineries, the twelve potentially impacted plants include three isolated asphalt refineries which appear to have no effective competition because of the high costs of transporting this specialty product. Therefore, although we can identify high water pollution control costs in twelve refineries, closer examination shows that no obvious force at work in the short or medium run would prevent any of the twelve refineries from passing their relatively higher costs on to consumers. We would expect no plant closings to result from the effluent limitations in this region.

The Midwest

Of the nine potentially impacted refineries in this region, three are asphalt producers with no asphalt competitors nearby. Three others are geographically isolated from all other competitors. These six refineries, then, will probably have sufficient market power to pass on relatively high abatement costs to their customers. The remaining three refineries face more difficult circumstances; they are situated among competing, lower-cost abaters within effective marketing distances. There is a clear possibility that these refiners will not be able to compete in their traditional markets.

The Gulf Coast

Among eighteen potentially impacted refineries in this region, eight are asphalt producers operating in relatively sheltered markets. Another refinery produces specialty lubricating oils, and another three refineries are geographically isolated or located only near other potentially impacted refineries. Thus, twelve refineries altogether appear to have no effective competitors with lower abatement costs. These refineries should not be expected to close, at least in the short or medium run, as a result of effluent limitations. Only the remaining six refineries in this region are situated among competitors who are, also, low-cost abaters. These six may be unable to compete effectively in their markets.

The Pacific Coast

Ten refineries in this region are potentially impacted. Of these, four are geographically isolated from their nearest competitors, and three produce asphalt in protected areas. The remaining three are situated among effective competitors with lower abatement costs. The market positions of these three are endangered by the effluent limitations.

In summary, twelve refineries at most (three in the Midwest, six in the South, and three in the West) are likely to be endangered by the promulgated effluent limitations. Taking account only of geographical isolation and markets for specialty products, we have identified 37 of the 49 potentially impacted refineries as bearing sufficient market power to pass on relatively high abatement costs to their customers, at least in the short run.

For a longer-run assessment of plant-level impacts, the foregoing conclusions must be modified. Many of the conditions that protect a class of potentially impacted refineries in the short run tend to diminish as time passes. Marginal refineries whose revenues fail to cover the *full* (fixed plus variable) costs of production and abatement are eventually depreciated and not replaced. Short-run barriers to competition disappear as refinery expansions and relocations come on-stream after the requisite planning time. Even the sources of economic rent may be eroded or depleted with the passage of time. For these reasons, we must conclude that our best estimate of the short-run plant-level impact of the 1972 amendments to the Act is an understatement of the long-run impact.

Long-run impacts will be further understated unless we recognize the dynamic role of new capacity expansion. All existing refineries compete in the long run with potential new or so-called "grassroots" refineries of latest design, and with incremental expansion of the most efficient existing refineries. Even refineries whose markets are not threatened by competition in the short run will face increasing competition as the existing stock of refineries is augmented and modernized.

Table 2-2 reports estimates of abatement costs for both grassroots refinery capacity and capacity assimilated by incremental expansion at existing sites.[8] New refinery capacity from both sources will apparently be able to comply with the regulations at lower cost than existing capacity. Average unit abatement costs constitute a burden of 9.1¢ per barrel on existing plants (computed at the 1983 abatement level); but only 6.9¢ per barrel for incremental capacity expansion at existing sites; and still less, 2.6¢ per barrel, for grassroots installations.

It is presumptuous to say that differentials in abatement costs will determine the future composition of capacity in the refining industry. These cost differentials are insignificantly small when compared to the broad range of issues considered by management when contemplating capacity expansion. For example, although a grassroots installation has an apparent 4¢ per barrel abatement cost advantage over incremental expansion at existing sites, the incentive to build a grassroots plant can be easily nullified by difficulties in obtaining a suitable site over the objection of local

Table 2-2
Unit Abatement Costs for New Capacity: Petroleum Refining Industry
(¢ per barrel)

	Abatement Level		
Type of Capacity	1977	1983	1985
Grassroots model 1[a]	1.8	3.1	3.3
Grassroots model 2[a]	1.7	2.5	3.1
Grassroots model 3[a]	1.7	2.4	3.0
Incremental expansion at existing sites	4.4	6.9	7.9
Average unit cost at existing plants[b]	5.2	9.1	10.8

[a]Three distinct new refinery models are treated. Each is of 250,000 bpd capacity. Models 1 and 2 are designed to accept high sulphur Middle-Eastern crude feedstock; Model 3 to process "sweet" domestic crude oil.

[b]This is the total cost to be incurred by existing facilities, reflecting expenditures already made by the industry, as well as those that are still impending.

Source: James L. Smith and Robert A. Leone, report to the National Commission on Water Quality, June 1, 1975, pp. 203-211.

environmentalists. The major usefulness of the cost estimates reported in Table 2-2 is to demonstrate the competitive threat that new capacity poses to older, less-efficient refineries.

Municipal Treatment of Refinery Wastewaters

Of the 250 U.S. refineries, 34 currently discharge wastewater to municipal or regional treatment facilities, rather than apply "in-house" treatment. This practice legitimately complies with the legislated effluent standards and is expected to continue through 1985. All our previous estimates of abatement costs reflected the appropriate municipal fees and charges for refineries using this service.

The significance of municipal treatment facilities is much greater for small refineries than for large. Economies of scale apply to almost all in-house treatment steps, conveying a competitive advantage to large refineries with the largest wastewater streams. Municipal treatment facilities pool the modest wastewater flows from a group of smaller manufacturing plants. This indirectly conveys the economies of scale back to the individual firms. Municipal treatment does for the small refineries what the economies of large-scale in-house treatment do for larger refineries.

To measure the benefit of municipal treatment, we have estimated the additional cost that would be incurred by users if they were forced to rely

Table 2-3

Abatement Cost Savings Resulting from Municipal Treatment: Petroleum Refining Industry

($ million)

Abatement Level	Capital Cost Saving	O & M Saving	Annual Saving
1977	44	11	18
1983	129	26	47
1985	164	37	63

Note: Savings are reported on a cumulative basis.

Source: James L. Smith and Robert A. Leone, report to the National Commission on Water Quality, June 1, 1975, pp. 212-221.

on expensive in-house treatment methods.[9] Our estimates (Table 2-3) indicate that the economies of scale provided by municipal treatment save the industry over $160 million in initial capital expenditures and more than $60 million in annual costs to achieve the 1985 standards.

The economic impact of pollution abatement on small refineries discharging to municipal facilities would surely be more severe if such facilities discontinued operation. Indeed, the magnitude of the savings resulting from municipal treatment suggests that many more small refineries may attempt to gain access to some form of cooperative treatment facility. For example, although none of the 49 highest-cost refineries examined earlier now use cooperative treatment, they have a strong incentive to do so in the future. If a trend to cooperative treatment does develop, the adverse economic impact on the industry will be mitigated, and our projected abatement costs will prove to be high estimates.

Sensitivity of Abatement Costs to Technological and Regulatory Influences

Within an individual refinery, the configuration of distinct process units largely determines the quantity and composition of water flows. A single refinery that subjects its throughput only to atmospheric distillation (a "topping" plant) generates a minimal waste load. A more complex refinery might subject the same barrel of throughput to repeated or consecutive processing states (e.g., distillation followed by various treating and finishing steps). Each successive stage of processing increases the quantity of water used and introduces additional pollutants to the wastewater stream. Both the American Petroleum Institute and the Environmental Protection

Administration have employed multiple regression techniques that explain the raw waste loads of a sample group of refineries as a function of process configuration and refinery size.[10] Clearly, the complexity of refinery operations influences the cost of water pollution abatement.

In view of the relationship between refinery complexity and abatement costs, it is important to identify current influences that may change complexity, and thus cost. Several prominent factors come to mind.

As a result of air pollution standards, refiners must treat crude feedstocks that have high sulphur and nitrogen content in order to remove these elements. Desulphurization essentially increases the intensity of refinery processing and introduces additional contaminants into the waste stream. Although the identity of major fields that will supply crude oil to the U.S. in the future is uncertain at this time, it is clear that if adequate oil supplies can be drawn mainly from the continental U.S., Canada, Alaska, or North Africa (all sources of low-sulphur crude oils), the cost of water pollution abatement in refineries will be mitigated. However, if supplies come increasingly from Venezuela and the Mideast (sources of high-sulphur crudes), as appears likely, then the abatement costs will be increased by the cost of clean-up in the desulphurization process.

The intensity of refinery processing is also influenced by the mix of demand for final products (e.g., gasoline, fuel oil, residual oil, etc.). Refiners have increased the amount of gasoline produced from a barrel of crude oil (now 50 percent of output) above the naturally occurring percentage (20 percent of each barrel) only by adding processing units which pollute the water. Raising the octane rating of gasoline similarly contributes to the pollution problem. Rising consumer demand for transportation fuels has mandated an historical rise in pollution-creating refinery processes.

Future trends in product mix are uncertain in the industry today. Recent gyrations in the price of petroleum products may have induced a "conservation ethic," thus changing the proportions of the traditional fuel demand. If, for example, the relative demand for gasoline were expected to decline (as a result of changing transportation modes), refiners would gradually revise their configuration of processes to less intensive patterns; the water pollution problem would decrease, and abatement costs would fall.

To illustrate the importance of changes in the product mix and the process configuration on the overall costs of water pollution control, we have reestimated abatement costs on the assumption that the complexity of refining operations is alternately higher and lower than is presently the case.[11] The estimates appear in Table 2-4.

Refinery complexity would increase (decrease) as illustrated if, for example, there were an addition (deletion) of cracking process units equal to 18 percent of feedstock capacity; or, an addition (deletion) of asphalt production units equal to 9 percent of feedstock capacity; or, an addition

Table 2-4

Increments to Abatement Costs Associated with Potential Changes in Petroleum Refinery Complexity

($ million)

	Increased Complexity		Decreased Complexity	
Abatement Level	Capital Cost	O & M Cost	Capital Cost	O & M Cost
1977	+21	+ 8	−25	− 8
1983	+98	+24	−110	−28
1985	+122	+37	−139	−42

Source: James L. Smith and Robert A. Leone, report to the National Commission on Water Quality, June 1, 1975, pp. 116-117.

(deletion) of lube process units equal to only 8 percent of feedstock capacity.[12]

We see that process changes of this size induce approximately a 7 percent variation in the capital costs of pollution abatement. Operating and maintenance (O & M) costs vary somewhat more, approximately 12 percent. Cumulatively, through 1985, capital costs might either rise by $122 million, or fall by $139 million, depending on whether refinery complexity were to increase or decrease. This range of variation translates into a 1¢-2¢ per barrel cost differential.

Given the sensitivity of pollution abatement costs to product mix and refinery complexity, it is interesting to discuss how public energy policy might influence the costs of abatement. Public policy can shape the mix of demand for refined products in several obvious ways. Gasoline, horsepower, or automobile-weight taxes are all measures which would reduce the proportion of refined gasoline sold relative to other petroleum products. In isolation, this should reduce the costs of water pollution abatement. However, federal air pollution standards, which necessitate the use of lead-free gasoline and catalytic converters in automobiles, exert the opposite influence. Alternative methods of producing high-octane gasoline without the benefit of tetraethyl lead invariably necessitate more intensive processing to restructure the hydrocarbon molecules. The government has also intervened directly to curtail demand for fuels with a high content of noxious elements. We have previously remarked on the cost of treating effluent from desulphurization units.

At this time, it is impossible to know which regulatory influence will most significantly influence the composition of final demand for petroleum products. However, in view of the demonstrated sensitivity of abatement costs to the complexity of refinery operations, we conclude that the actions of federal and state government will help shape the economic impacts resulting from water pollution abatement efforts.

Summary

Few petroleum refineries will be impacted significantly by the 1972 pollution abatement legislation. The average level of unit abatement costs is low and spread uniformly across the industry. Abatement costs constitute substantially less than 1 percent of the final price of refined petroleum products. The structure of the industry is basically competitive and conducive to a pass-through of abatement costs to ultimate consumers, even though the industry may have excess refining capacity and thus experience intense price competition while effluent regulations go into effect. In the event of the hypothesized cost pass-through, the strength of consumer demand for petroleum products appears to be sufficient to absorb nearly all abatement costs.

Approximately 50 very small refineries (2 percent of capacity) operating throughout the U.S. experience uncommonly high unit abatement costs. Our analysis suggests that the number of plant closings induced by effluent limitations will be small, even in this group. Potential closings are likely to be confined to a subset of twelve refineries put at an obvious competitive disadvantage by the legislation. However, there are compelling arguments which suggest that few even of these twelve will leave the refining industry in the near future.

The magnitude of abatement costs is even less significant relative to the size and volatility of other disturbances afflicting the refining industry. Concern over the cost and availability of crude oil feedstocks dominates industry management. Anticipation of and reaction to miscellaneous government regulations (e.g., price controls, import quotas, auto emission standards, siting regulations, etc.) occupy still more of management's attention. It is foolish to believe that the imposition of a very marginal level of water pollution abatement costs will profoundly impact the industry. Abatement expenditures will be incurred, prices will rise (although the increment may not be discernable against the background pattern of price variations), and the industry will continue on about its business, much as before effluent limitations.

Finally, the absence of highly significant and visible impacts should not lead one to overlook the sensitivity of real abatement costs to a multitude of factors under both government and industry control. We have identified several areas and trends that promise to shape the distribution and level of abatement costs, even if not affecting their order of magnitude. A comprehensive assessment of abatement costs and impacts in the refining industry requires that these rather volatile influences be monitored carefully, by both federal regulators and industry management.

Notes

1. *Oil and Gas Journal*, July 1, 1974.

2. A detailed derivation of all the cost estimates used in this chapter is found in James L. Smith and Robert A. Leone, *The Economic Impact of the 1972 Amendments to the Federal Water Pollution Control Act on the Petroleum Refining Industry*, report to the National Commission for Water Quality, National Bureau of Economic Research, June 1, 1975.

3. Based on capacity of existing plant and equipment equal to 15.141 million barrels per stream day, reported in *Oil and Gas Journal*, July 1, 1974. We assume 347 "stream days" per year.

4. These calculations and projections are reported in more detail in Smith and Leone, op. cit., pp. 168-175.

5. *Oil and Gas Journal*, August 12, 1974.

6. *Project Independence Report*, Appendix A-II, p. 60.

7. These projections are outlined in Smith and Leone, op. cit., p. 168-178.

8. Derivation of the estimates is discussed in Smith and Leone, op. cit., pp. 204-211.

9. The derivation of these cost estimates is discussed in Smith and Leone, op. cit., pp. 212-220.

10. These methods are described more completely in: Brown and Root, Inc., "Analysis of the 1972 API-EPA Raw Waste Load Survey Data," American Petroleum Institute, Publication Number 4200, Washington, D.C., (July 1974); U.S. Environmental Protection Administration, "Development Document for Effluent Limitations Guidelines and New Source Performance Standards for the Petroleum Refining Point Source Category," Washington, D.C., (April 1974).

11. Explanation of the estimation technique is found in Smith and Leone, op. cit., pp. 112-121.

12. Smith and Leone, op. cit., p. 119.

3

The Pulp and Paper Industry

Robert A. Leone,
Richard Startz,
and Mark Farber

The pulp and paper industry is one of the largest industrial users of water and, consequently, has always had a major effluent control responsibility. In this chapter we develop estimates of the impact of the 1972 water pollution controls on this industry using a simple computer simulation model which permits us to focus on problems of short-run industry dynamics.

Cost Estimation Methodology

At the core of our study of water pollution control impact lies the estimation of effluent control costs. Although there are numerous determinants of the abatement costs that will be experienced at any given mill, the volume of effluent flow is the only major determinant for which we have enough individual mill observations to permit us to calculate the *MCC* curve. For each of the 392 mills considered here, we have extrapolated abatement costs appropriate to its level of effluent flow, thus deriving the capital cost and an annual operating and maintenance cost necessary to bring each existing mill into compliance with each of the three successive treatment levels mandated by the 1972 water pollution control legislation.[1] For convenience, we report these costs on an "annual unit cost basis"; that is, we report the number of dollars necessary annually to achieve abatement per ton of capacity. For the purpose of cost estimation, we assume that capital equipment is discounted at a 10 percent annual rate over a useful life of 10 years or, equivalently, we assume an annual "capital recovery factor" of 0.163. The annual unit cost for a plant is then the operating and maintenance cost for effluent controls plus 0.163 times the capital cost of effluent controls, all divided by tons of (annual) capacity.[a]

The assignment of abatement costs to each of 392 mills raises important questions regarding the introduction of internal process changes as substitutes for end-of-pipe waste treatment. Since effluent treatment costs de-

---

[a] Although Chapter 1 identified four different cost concepts, in this chapter we restrict the discussion to "impending costs." A detailed presentation of the other cost estimates is found in Leone, Startz, and Farber, op. cit., Section II.

Table 3-1
Impending Costs of Water Pollution Control for the Pulp and Paper Industry by Abatement Level *(1973 prices)*

	Cumulative Costs		
Treatment Level	Capital ($ million)	O & M ($ million/year)	Average Annualized Cost ($/ton)
BPT	1929	99	3.33
BAT	2539	145	4.50
1985 Goal	3855	229	6.90

Note: A ton of pulp sold for approximately $180 in 1973.

Source: Robert A. Leone, Richard Startz, and Mark Farber, report to the National Commission on Water Quality, June 1975, Exhibits 26, 28, and 29.

pend primarily on the volume of water that carries the pollutants, a plant could introduce flow-reducing production changes that would require a much smaller treatment plant rather than construct a facility capable of treating current flow levels. Typically, such a change requires a capital expenditure which is partially or fully offset by savings in ongoing operating and maintenance costs. Our estimates of cost reflect these possible "internal changes" wherever they are economically justifiable.[b]

Table 3-1 presents an estimate of water pollution costs aggregated over the entire pulp and paper industry. The costs are cumulative, meaning those reported as necessary to achieve 1983 effluent limitation requirements include the costs required to comply with the 1977 standards. Production cost savings and capital costs associated with internal changes we estimate to be economically justifiable in the absence of the Act are excluded.

As noted in Chapter 1, the industry's marginal control cost curves are potentially more interesting than the aggregate costs of wastewater treatment. Figure 3-1 shows these *MCC* curves for the 1977, 1983, and 1985 abatement levels.

Although the average unit treatment cost in the industry needed to achieve the 1977 abatement level is $3.33 per ton,[c] Figure 3-1 shows that 25 percent of capacity will incur costs of $1.40 or less and up to 50 percent can achieve the mandated effluent reductions for $2.40 or less. Some plants, however, have much higher costs. The mill at the 93rd percentile will incur costs of $8.40 per ton unless it closes down. The mill at the 99th percentile has costs of $14.50 per ton. As the nearly vertical line at the right of the chart shows, the plants in the worst position are very badly off

[b] To be economically justified, an internal process change must result in sufficient operating cost reductions to yield a 10 percent return on the required capital investment.

[c] In 1973, a ton of pulp sold for approximately $180.

Annualized Unit Cost ($/ton)

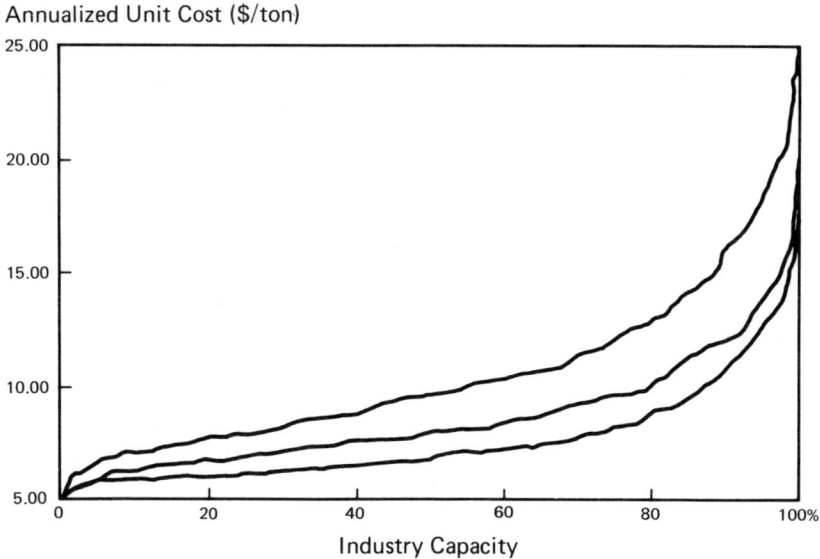

Notes: Upper curve: 1985 (EDOP) Abatement levels:
                 Middle curve: 1983 (BAT) Abatement levels.
                 Lower curve: 1977 (BPT) Abatement levels.

Source: Robert A. Leone, Richard Startz, and Mark Farber, report to the National Commission on Water Quality, June 1975, Exhibit 37.

Figure 3-1. Marginal Control Cost (*MCC*) Curves: Pulp and Paper Industry

indeed. We estimate the last 1 percent of capacity will have costs ranging from $14 to $75 per ton. The likely consequences are obvious. The large number of plants that have relatively low clean-up costs will not be badly hurt (and, as pointed out in the first chapter, they may even profit), by the imposition of stringent water pollution controls. Those plants with the highest costs are probably too expensive to even consider remaining in operation.

Cost Sensitivity to Input Factor Prices

In assessing the reasonableness of any of these cost estimates, we must acknowledge that changes in factor prices may affect not only the costs of pollution abatement, but also the treatment technology chosen. Land prices illustrate the importance of this phenomenon. To comply with the

1977 (BPT) standards, two principal technological options exist: one utilizes an activated sludge (AS) treatment system and the other employs an aerated stabilization basin (ASB). AS has high nonland costs but requires little space. ASB has lower nonland costs but requires large amounts of land. It is often felt that purchasing land for ASB makes it "prohibitively expensive." To determine the validity of this contention we calculated the break-even cost of land for integrated mills which would make them indifferent to the two options. Our analysis concludes that on average, mills will choose ASB as long as land costs are less than $227,000 per acre, a rather considerable sum indeed. Our engineering cost data suggest that every mill would choose ASB unless there were some noneconomic barrier, physical or legal, to obtaining it.

The usefulness of this simple break-even calculation is obvious. The engineering costs available to us indicate that virtually all mills will choose ASB. There are very few plants where land is not available—or cannot be created!—for $227,000 per acre; yet in practice, there are plants using AS treatment. Available data do not permit us to determine whether these plants have unique characteristics which justify this choice. In any case, the high break-even costs lead one to question whether the *relative* costs of AS vs. ASB have been accurately measured.[d] Interestingly, we observe similarly high break-even land-cost factors in other industries where, there too, the conventional wisdom has it that option choice is sensitive to the price of land. Surprisingly, the simple break-even calculations are not typically performed; yet they can easily quantify the rather imprecise concept of a "prohibitive expense."

Lacking an economically defensible alternative, we assume that all mills do choose ASB. Except where indicated below, we arbitrarily assign a land cost of $3,000 per acre for integrated mills (that is, paper-making facilities with a pulp-making capacity) and $20,000 per acre for nonintegrated mills (often found in higher-density locales). Since it is impractical to discover the actual land costs at each mill, and since the price of land may well rise as soon as mills start to bid for it, we would like a sensitivity analysis showing how the price of land will change the marginal cost curve. To illustrate how sensitive the *MCC* curve is to land prices we assumed that all mills would have to pay the clearly absurd sum of $180,000 per acre for land and recalculated the marginal cost curve. Typically, costs increase by perhaps 10 percent. This indicates that if the price of land should skyrocket there would be a serious, but not catastrophic, effect on abatement costs. As usual, those mills that already have the highest costs are hit worst. For the high-cost 10 percent of the mills, abatement costs rise by as much as 20 percent using the hypothetical land price of $180,000 per acre.

---

[d] We must stress that these calculations raise doubts only about the *relative* costs of these two options. The absolute cost estimates may not be very far off.

Annualized Unit Cost ($/ton)

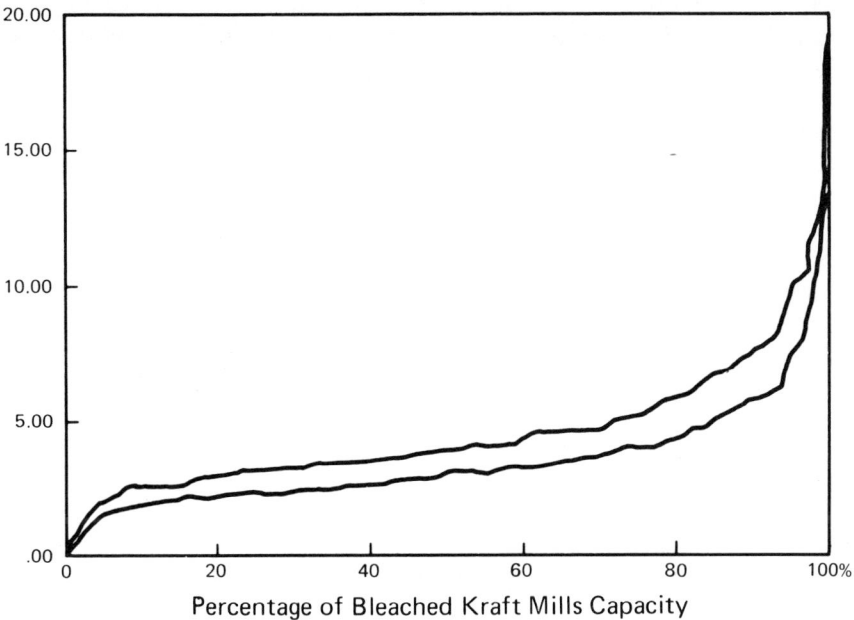

Percentage of Bleached Kraft Mills Capacity

Notes: Upper curve reflects costs at current brightness levels.
 Lower curve reflects costs after a 10% reduction in brightness.

Source: Robert A. Leone, Richard Startz, and Mark Farber, report to the National Commission on Water Quality, June 1975, Exhibit 57.

Figure 3-2. Effects of a Reduction in the Bleaching Level in Bleached Kraft Mills on the Marginal Control Cost Curve

Cost Sensitivity to Changes in Product Mix

Pulp must be bleached to turn it from the color of brown kraft shopping bags into whiter shades for writing papers and other uses. Because the bleaching process produces much effluent it provides a fine example of a possible product change that can reduce the costs of complying with water pollution requirements. A small reduction in the amount of bleaching may result in a large saving in treatment costs. We estimate that a reduction in the brightness of paper from a level of 88 to 80 on the industry's brightness scale would reduce by approximately one-third the wastewater flow associated with the bleaching operation.[e] Since bleaching accounts for approximately

[e] This flow reduction is *after* all economically justified internal process changes that can also reduce flow have been made. We thank Professor William McKean of North Carolina State University for providing us with data to make these calculations.

one-half the flow in a bleached kraft mill, overall flow would fall by about one-sixth. Figure 3-2 shows the *MCC* curve for bleached kraft mills with and without this hypothetical brightness reduction. We estimate that this relatively modest product change would yield on average a 24 percent savings in abatement costs. Whether the market will accept such a change is open to question. Nonetheless, it seems likely that a bleaching reduction will be a viable option for some mills, especially those facing the highest costs.

Industry-Level Impact

As Chapter 1 explains, we are primarily interested in the effects of water pollution controls on prices, production levels, and profits. We have chosen to estimate these impacts for the pulp and paper industry using an econometric model. Although the model is simple in its construction, it does attempt to capture the important dynamic consequences of effluent abatement.

We formulate the model using equations to reflect supply, demand, investment, and capacity relationships. The model is then modified to reflect the advent of water pollution regulation. By using the computer to simulate both sets of equations, we are able to chart the difference between the two situations and thus simulate the impact of the Act. Before delving into the specifics of the model it seems wise to emphasize the model's limitations. First, the model is limited in its accuracy by its simple design and second, by the rather wide confidence intervals on some of the econometrically estimated input parameters. In addition, the model is confined to the production of pulp.

If the basic aim of our model is to capture a price differential, the basic obstacle must be that we do not have any information on production costs. Without such data (which are considered proprietary), it is difficult to describe the supply curve. Without the supply curve, it is difficult to estimate statistically the demand curve. And without cost data we cannot confidently estimate what part of treatment costs the industry will be able to absorb, and which they will pass through to customers. Our model is specifically built to avoid problems associated with these data deficiencies.

The basic model relationships are:

1. Short-run supply (Q) is a function of current capacity (C).

$$Q = Q_s(C)$$

2. Demand (Q) is a function of the price of pulp (P) and the level of national aggregate income (Y).

$$Q = Q_d(P, Y)$$

3. Investment (*I*) is a function of the price of pulp and the production cost of pulp in new mills (*COST*).

$$I = I(P, COST)$$

4. Current capacity is a function of capacity in the prior period (C_{-1}) and recent investment levels (I, I_{-1}, I_{-2}).

$$C = C(C_{-1}, I, I_{-1}, I_{-2})$$

To determine the impact of imposing effluent controls the above four equations must be adjusted to account for the plant closings due to the Act as well as the capital and operating costs associated with the clean-up of both new and existing capacity. We consider below each of these four equations as well as the additional equations required to account for the impact of water pollution controls.

Supply

Although one might plausibly argue that in truth the pulp industry is mildly oligopolistic, we assume that we are modeling a competitive industry. We have also chosen to assume that the short-run supply curve is vertical. We have made this assumption because important dynamic impacts occur principally when capacity is a bottleneck (e.g., when the supply curve is in an essentially vertical position), hence, the phenomena we wish to observe can only occur when the rate of capacity utilization is high. This assumption principally reflects the lack of production cost data to support an alternative. In the past, the industry has in fact experienced sustained periods of close to full capacity utilization indicating that this is not an unreasonable assumption. Water pollution controls themselves encourage such high rates of utilization both because of possible reductions in total industry capacity (for example, plant closings), and because the capital intensity of the abatement options encourage full utilization. In any event, we employ an equation of the form

$$Q = 0.96C \tag{3.1}$$

where 0.96 is the highest historically observed rate of capacity utilization.

Demand

Because of the deficiencies of publicly available data, it was not possible to take full account of several important factors in estimating the demand equation. Primary among these are the possibilities for substitution of other products for pulp and the use of the supply curve to properly identify the

Table 3-2
Demand Parameters

	A_0	A_1	A_2
coefficient	9.45	−2.33	0.66
standard error	7.07	1.22	0.40
$R^2 = 0.93$		standard error of regression = 0.05	
degrees of freedom = 16		Durbin-Watson = 1.79	

demand curve. We estimated the equation in log form using a constant elasticity specification. We also assumed that the proper specification was in real, per capita measures.

$$\log(Q/POP) = A_0 + A_1 \log P + A_2 \log(Y73/POP) \qquad (3.2)$$

where *POP* is population and *Y73* is GNP in 1973 constant dollars.[f] We estimated the equation over the time period 1955 to 1973 using annual data. The log of capacity and a time trend representing technological change were considered the predetermined variables from the supply equation. Table 3-2 shows the results of using an instrumental variable estimation method.

There are at least two defects in the equation. First, the estimated price elasticity of −2.33 is far higher than anyone has previously suggested. One cause may be that this is really a long-run elasticity while our analysis calls for a short-run elasticity. Second, there are very large standard errors relative to the size of the estimated coefficients. This may in part explain the first difficulty. Our response to both of these is to recognize that we can place little confidence in our estimate of the price elasticity. Consequently, we perform simulations under three alternative assumptions of elasticity: −2.33, −1.0, and −0.5.

Investment

Unfortunately, investment equations are the bane of econometric modeling. To compound problems, we require an equation which would be sensitive to an exogenous reduction in capacity associated with legislatively induced plant closings. We formulated the following specification in which producers respond, in a partial adjustment manner, whenever the price rises above the long-term cost.

$$I/C_{-1} = C_0 + C_1 \log P + C_2 T \qquad (3.3)$$

[f]All prices in this chapter, unless stated otherwise, are in 1973 constant dollars.

Table 3-3
Investment Parameters

	C_0	C_1	C_2
coefficient	−1.59	26.7	0.36
standard error	73	12.6	0.12
$R^2 = 0.75$	standard error of regression = 43		
degrees of freedom = 12			

Table 3-4
Capacity Parameters

	B_0	B_1	B_2	B_3
coefficient	0.987	0.00424	0.00513	0.00258
standard error	0.028	0.00312	0.00287	0.00267
$R^2 = 0.98$		standard error of regression = 0.45		
degrees of freedom = 11		autocorrelation coefficient = 0.75		

where I is real investment and T is a time trend representing technological change. Because ordinary least squares produced a Durbin-Watson of 2.40 the estimate in Table 3-3 was done using the Cochran-Orcutt iterative technique. The estimated coefficient of serial correlation was −0.29.

Capacity

The capacity equation in the absence of the Act is straightforward. The equation is corrected for autocorrelation. (See Table 3-4).

$$C = B_0 C_{-1} + B_1 I + B_2 I_{-1} + B_3 I_{-2} \tag{3.4}$$

Impact Equations

In Chapter 1 we noted that pollution regulations not only raise the supply curve but simultaneously shift it to the left. Since we assume a vertical supply curve, any upward shift becomes irrelevant for our purposes. The shift to left, however, remains crucial to our analysis.

This shift can result either from some plants closing as a result of the Act or fewer new plants opening. There may be fewer new plants, since pollution abatement raises costs and, therefore prices, and in the long run, the demand for pulp will fall. Thus, less capacity will be required. Also, in the short run, capital may be "diverted" from investment in new capacity to pay for cleaning up existing plants. If capital markets were perfect, this diverting would not occur, since the profitability of investments in new plants is not diminished, *ceteris paribus,* by additional investment being required in old plants. Nonetheless, in the less-than-perfect real world this may be a significant effect. Having no real way to decide to what degree investment in new plants will be displaced, we simulated our model using two alternative extreme assumptions. Under an assumption we call "Type 1" investment, all the capital expenses associated with wastewater treatment are subtracted from normal investment in the industry. In what we call "Type 2" investment, no such reduction is made; in this alternative extreme, all water pollution related investments are net additions to the normal investment of the industry.

We also made one additional adjustment. Part of the gross investment in new plants will be diverted to pay for pollution control in the new plants themselves. To estimate the fraction of gross investment that will actually go into production equipment, we first assumed that any new plants that are constructed will experience effluent control costs per unit output equal to those experienced by the lowest-cost 5 percent of current capacity. We further assume that the capital investment required to construct a ton of pulp-making capacity is $400 and all new plants will be built to meet 1983 standards.[2] These assumptions indicated that 96.1 percent of gross investment will actually go into pulp production equipment. This may be a low estimate, since a great deal of flexibility exists before construction begins. This flexibility often permits pollution standards to be met at surprisingly low costs in new facilities. However, we might stress that our results are not particularly sensitive to the fraction of gross investment that will actually go into production equipment.

Using the price change to close the model—that is, to equate supply and demand—and produce the required capacity reductions (plant closings), we derive the net investment and capacity equations:

$$\text{Type 1 investment: } NI1 = 0.961(I - CLEAN) \qquad (3.5)$$

$$\text{Type 2 investment: } NI2 = 0.961I \qquad (3.6)$$

where *CLEAN* is the capital cost to provide existing facilities with an effluent abatement capability. In either case

$$NC = C(NC_{-1}, NI, NI_{-1}, NI_{-2}) - CAPOFF \qquad (3.7)$$

that is, net capacity (*NC*) after the Act is a function of net capacity in the prior period (NC_{-1}) and recent levels of investment (NI, NI_{-1}, NI_{-2}) and plant closings ($CAPOFF$).

Since we have no production cost data, we can make only a limited approximation to the change in profit. We assume that no quasi-rents exist anywhere in the industry prior to the implementation of more stringent water pollution controls. Specifically, we assume that the long-run marginal cost is a constant equal to the observed price before controls. Under this assumption, the net revenue after abatement expense can be written

Type 1 investment:

$$CASH1 = (P' - P) \cdot Q - OM \cdot OLDCAP - OMNEW \cdot NEWCAP$$
(3.8)

where OM is the operating and maintenance (O&M) cost for effluent control in existing facilities ($OLDCAP$), and $OMNEW$ the O&M cost for abatement in new facilities ($NEWCAP$). P and P' are the prices of pulp before and after the Act, respectively.

Type 2 investment:

$$CASH2 = (P' - P) \cdot Q - OM \cdot OLDCAP - OMNEW \cdot NEWCAP$$
$$- CLEAN \qquad (3.9)$$

$OLDCAP$ is adjusted both for plant closings ($CAPOFF$) and normal depreciation. $OMNEW$ was arrived at by assuming the cost to be the same as that experienced by the lowest-cost 5 percent of existing plants. As our measure of profits seems more like a cash flow measure, we have labeled it $CASH$.

The question of depreciation is quite difficult to treat rigorously. The coefficient B_0 (0.987) from equation (3.4) is one measure of depreciation, reflecting approximately a 2 percent annual depreciation rate. The application of this parameter to total capacity in the prior period implicitly assumes that depreciation comes totally from individual machines wearing out within plants. If depreciation in the industry takes the form of closing plants, then our estimate of capital spending on abatement equipment will prove high, since the most costly plants to clean up are also among the oldest in the industry. Since B_0 is quite close to 1.0, small variations in this parameter have no great practical importance.

It is necessary to decide in the dynamic model when the various abatement expenses are incurred. We assumed that the capital costs for 1977 standards would be borne 20 percent in 1974, 20 percent in 1975, and 60 percent in 1976. We assumed the same percentages in 1980, 1981, and 1982 respectively for 1983 standards. For the 1985 level, we assumed 40 percent in 1983 and 60 percent in 1984. These cost incidence assumptions reflect the fact that initial planning stages in a capital project often involve very little money. The largest sums are typically expended at the time of construction. A shift in the time incidence might significantly affect the results of our analysis. Certainly both the rigor of enforcement and the speed of any resulting litigation will change the timing of investment and consequently, the discounted costs to the industry of the Act.

The actual equations of our computer simulation model are reported in Appendix 3A to this chapter.

Before presenting any simulation results, we will discuss further some of the assumptions underlying the model.

Making Some Assumptions More Explicit

The position of the demand curve is important to our analysis, but except for using alternative demand elasticities, we have passed over it rather lightly. We assume that demand depends directly on national income. We then assume national income will grow smoothly. While this is obviously not a correct assumption, we would be rather foolhardy to attempt to predict business cycles as part of our model. Nonetheless, if a slump happened to occur at the same time as a compliance date, this might well mitigate capacity shortfalls associated with plant closings and the diversion of investment dollars and, therefore, reduce or eliminate the level of simulated price increases. It might, of course, also reduce investment somewhat. Further, we must clearly separate a measure of impact based on current prices and one based on a price level that would have been observed in the absence of the Act. If prices would have fallen 10 percent in the absence of controls, but instead stay steady, some might say the impact of the Act had been eliminated; we would say there has been an 11 percent price increase due to the imposition of water pollution controls.

The very assumption that pulp demand depends rather directly on GNP is open to question. Recent history seems to support this generally accepted assumption and we do not doubt its basic accuracy over any moderately long period of time. However, in all but the most recent past we have experienced a relatively stable market for pulp and paper. The very events we are studying will upset the market and it is easy to believe that there will be short-term fluctuations we have not taken into account.[3]

This is not a question of idle curiosity. In 1974 the price of pulp rose dramatically and our model, which focuses on supply-side influences, fails to predict so substantial an increase. Of course, one could attribute the 1973-1974 price experience in the pulp industry primarily to demand-side influences; for example, a sudden, temporary increase in speculative demand for pulp associated with world-wide commodity price inflation, the elimination of price controls and other factors. Such an aberration in demand is outside the purview of our model and is one of the many things we do not attempt to take into account.

In the process of closing the model by comparing the cost increase to the price increase and thus determining capacity reductions, we encounter several difficulties.

The first problem is one of static analysis attributable to our lack of production cost data. As we noted in Chapter 1 (Figure 1-5), the capacity reduction due to pollution regulations depends on the difference between variable cost and total cost in the region of the intersection of the demand curve with the postabatement cost curves. Clearly, we do not know the shape of either the total or the variable cost curve and, therefore, must assume they are both flat (as we did in drawing Figure 1-5). However, what is far more detrimental is that we have no idea as to the *difference* between total and variable costs. The only assumption we can make is that total cost and variable cost are everywhere equal to each other and that both equal price. This is exactly equivalent to assuming that producers can pass on all abatement costs, excepting that those with costs greater than acceptable in the market will close. This clearly overestimates the price increase since it suggests that more plants will close than is in fact the case. It also underestimates losses, since it effectively assumes there are none. The one redeeming characteristic of this assumption is that for small indicated reductions in capacity, the plants which close may indeed have variable costs very close to total cost. Essentially it appears that those plants high up on the *MCC* curve are generally also the oldest and smallest plants and, therefore, may well have already fully depreciated their fixed capital investment. As a result of these observations, we know that a simulation that indicates just a few closings overstates the impact, but such overstatement may be small. On the other hand, if a simulation predicts a substantial number of closings, we are outside the region in which we can reasonably expect total and variable costs to be equal. In this case, we can only say that the results are not reliable.

The static analysis ignores the fact that management does not face a single cost or price but rather a continuous stream of each over a period spanning almost two decades. This time stream of costs and prices presents a second problem to the model builder, though not one quite so severe as it presents to industry management. Unfortunately, we have embarrassingly little advice for managements facing this timing problem. We can, however, offer the modeler two solutions. One is to estimate plant closings using a rational expectations framework and the other is to make conditional predictions of impact, based upon a priori specifications of plant closings (*CAPOFF*). We chose to do both.

To estimate the level of plant closings, we assume that the pulp industry has "rational expectations" about future price levels—that is, if acted upon, the expectations will be fulfilled. Thus, we want to find the level of plant closings such that the price increase resulting from those closings would just cover the abatement costs of the marginal plant. To do this we compared the present value (at a 10 percent discount rate) of abatement costs (employing the incidence assumptions mentioned above) to the present value of the simulated revenue stream of each mill (through 1990, the end of the simulation). (However, we set any water pollution induced price rise before 1977 to zero on the grounds that a plant can receive the benefit of such an increase without meeting the standards.) For any given set of assumptions about demand and investment, the level of plant closings which causes the discounted cost stream to equal the discounted revenue stream, subject to the previous caveat about total and variable cost, is said to follow from an assumption of rational expectations.

It is not at all unreasonable to take the rational-expectations hypothesis with a grain of salt; in particular, we have not explained how, precisely, the industry is supposed to deduce the course of the market over the next twenty years. If the reader is uncomfortable with this assumption (as are the authors) it is possible to consider the results of our simulations as conditional predictions. Thus we say simply that *if* 5 percent of capacity is closed in 1977, *then* (subject to all the caveats that have gone before) prices will rise by the simulated amount.

Testing the Model

Figure 3-3 shows simulation results for 1960 to 1973 with no water pollution expenditures. Note that the test was run assuming a utilization rate of 96 percent; although this assumption may be relevant for our impact analysis, it clearly weakens the model's ability to replicate the past. It should also be noted that Figure 3-3 and all graphs that follow must be read with care since the scales on the axes have been chosen to magnify the variation for the sake of visibility. We readily acknowledge that a strong correlation between actual and simulated performance when based on annual time series data is not a very strong test of a model.

Table 3-5 shows the simulated annualized pulp prices in 1973 dollars, for three different price elasticities, both investment assumptions, and three different capacity cutoff levels.

Keeping in mind that the base price of pulp in 1973 was approximately $180 per ton, we can examine the sensitivity of the model to the different assumptions. The extreme predictions, $1.54 to $18.96, cover a range of only 10 percent of the base price of pulp. Further, holding any two of the assumptions constant, the maximum range is less than $12.00. There are at least two possible interpretations of these findings. One is that the rela-

Price ($/ton)

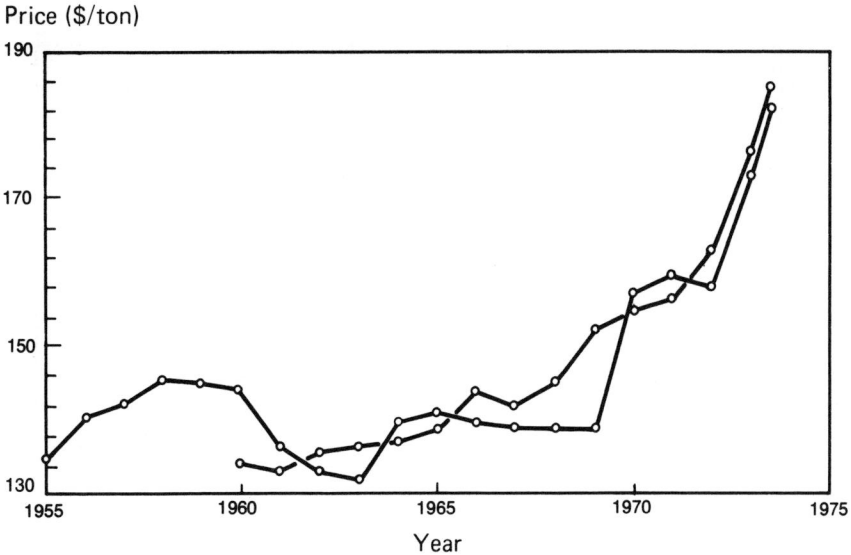

Notes: Longer curve reflects actual current dollar prices per ton pulp.

Shorter curve reflects predicted current dollar prices per ton pulp assuming a price elasticity of −2.33, in the absence of abatement impact.

Source: Robert A. Leone, Richard Startz, and Mark Farber, report to the National Commission on Water Quality, June 1975, Exhibit 65.

Figure 3-3. Actual and Predicted Current Prices for Pulp—Test Simulation

Table 3-5
Summary Simulation Results: Pulp Industry Impact Model

Assumed Capacity Reduction Due to Abatement Costs (%)	Annualized Cost of Abatement ($/ton)	Investment Type	Annualized Revenue[a] ($/ton)		
			Price Elasticity		
			−2.33	−1.0	−0.5
5	13.94	1	10.12	13.54	14.20
		2	1.54	2.16	2.45
15	8.99	1	10.63	14.56	15.96
		2	3.60	5.15	6.04
25	6.87	1	11.91	16.64	18.96
		2	5.83	8.44	10.14

[a]Note that this is the *annualized* equivalent of the simulated price time stream. In any given simulation year, predicted price increases can be higher or lower than this annualized equivalent.

Source: Robert A. Leone, Richard Startz, and Mark Farber, report to the National Commission on Water Quality, June 1975, Exhibit 73.

Price Increase (%)

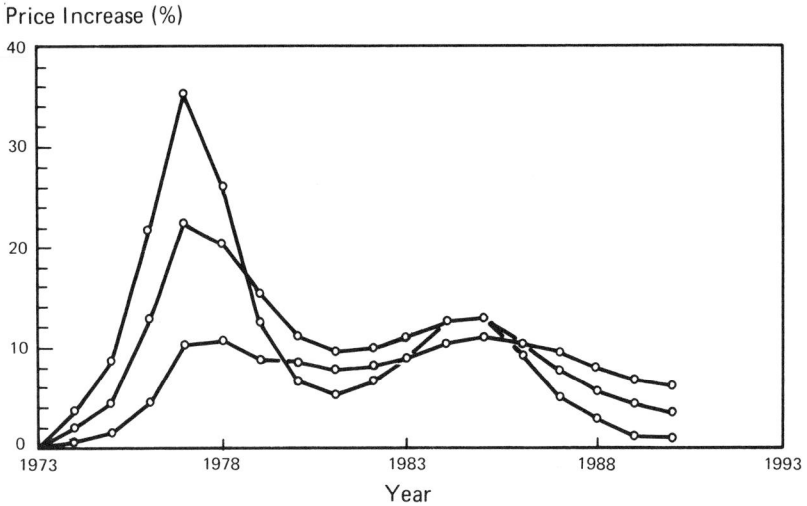

Notes: Upper curve: elasticity = −0.5.
          Middle curve: elasticity = −1.0.
          Lower curve: elasticity = −2.33.

Source: Robert A. Leone, Richard Startz, and Mark Farber, report to the National Commission on Water Quality, June 1975, Section III, Appendix, p. 248.

Figure 3-4. Simulated Percentage Increases in Pulp Prices Due to Effluent Control Over Time for Alternative Demand Elasticities Given Type 1 Investment

tively narrow range of our predictions reassures us about the usefulness of the model. A less sympathetic interpretation is that since the predicted increases are such a small portion of price, these added costs of control may very well be covered by the difference between total and variable costs. In this case, no closings at all would be expected in the short run.

The revenue changes in Table 3-5 are *annualized equivalents* of the simulated price stream over time; hence, for any specified point in time, the model will simulate a price higher or lower than this annualized equivalent of the cost and revenue as indicated in the table.

Figure 3-4 shows the simulated price time stream for all three different elasticities, assuming Type 1 investment and a capacity reduction of 5 percent. Note that the lower the elasticity, the greater the short-run price "blip," with the peak price increase approaching 40 percent. Figure 3-5 shows the capacity difference resulting from the two different investment assumptions. The range is considerable, indicating that we clearly require

Change in Pulp Capacity (%)

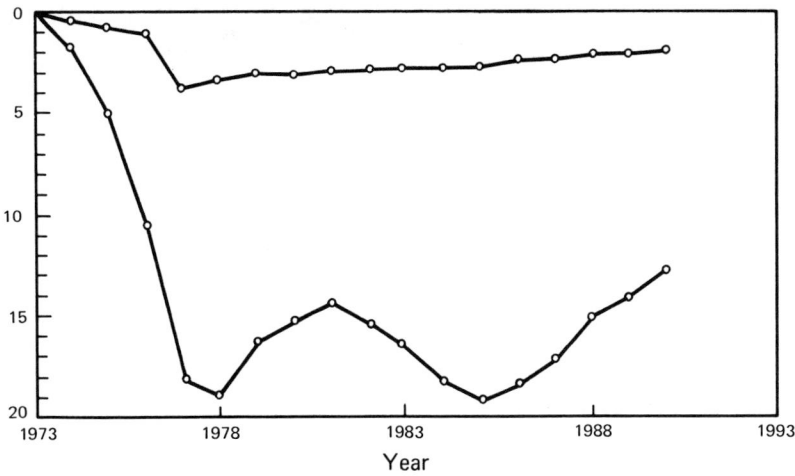

Notes: demand elasticity of −2.33 and a 5 percent capacity reduction.
      Upper curve reflects Type 2 investment.
      Lower curve reflects Type 1 investment.

Source: Robert A. Leone, Richard Startz, and Mark Farber, report to the National Commission on Water Quality, June 1975, Section III, Appendix, p. 273.

Figure 3-5. Simulated Percentage Changes in Pulp Capacity Due to Effluent Control Over Time for Type 1 and Type 2 Investment

more empirical evidence on investment behavior in the pulp industry to determine where reality lies between these two simulated extremes.

    Figure 3-6 is the cash flow diagram. The range between the two investment assumptions is over $1 billion, indicating that the prediction range of our model for profits is too wide to be useful. This is far from surprising given the compromising assumptions we made due to the lack of production cost data.

Summary

In the long run, our analysis suggests that prices in the pulp and paper industry are likely to increase about 4 percent due to stringent water pollution controls, reflecting our estimate of the added costs of water pollution control in new mills. Technological change, factor substitution, and changes in the product mix over time may lower even this small cost

Cash Flow ($ million)

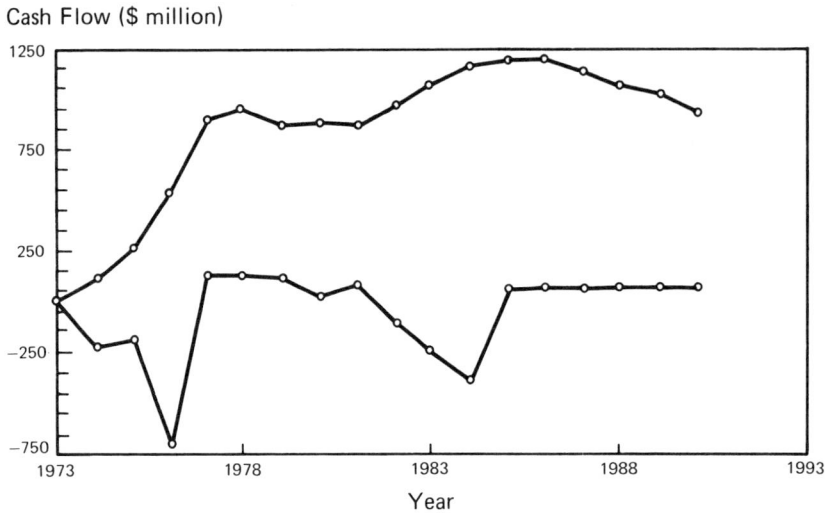

Notes: demand elasticity of −2.33 and a 5 percent capacity reduction.
      Upper curve reflects Type 1 investment.
      Lower curve reflects Type 2 investment.

Source: Robert A. Leone, Richard Startz, and Mark Farber, report to the National Commission on Water Quality, June 1975, Section III, Appendix, p. 257.

Figure 3-6. Simulated Cash Flow for Type 1 and Type 2 Investment: Pulp and Paper Industry

increase substantially since many cost-reducing opportunities appear to exist in the industry (e.g., internal changes and reduced brightness). It remains to be seen, however, if the requirements to maintain a flexible operating capability and the dictates of the market will permit these cost savings to be realized.

In the short run, as the simulation exercise indicated, few quantitatively precise conclusions are possible. Depending on prevailing aggregate economic conditions and the investment behavior of the pulp and paper industry, short-run price increases of 5 to 10 percent attributable to pollution controls would certainly not be unreasonable. Substantially higher increases also appear to be possible if there were to prevail certain plausible conditions which might accurately describe the pulp and paper industry.

Perhaps our final conclusion is that more precise empirical assessments of the impact of water pollution controls on the pulp and paper industry must await further methodological and data collection advances.

Notes

1. Robert A. Leone, Richard Startz, and Mark Farber, *The Economic Impact of the 1972 Amendments to the Federal Pollution Control Act on the Pulp and Paper Industry,* report to the National Commission on Water Quality, National Bureau of Economic Research, June 1975.

2. Bruce R. Lippke, Gerald M. Hughes, and John A. Carrougher, "The Impact of Pollution Standards on Shortages, Inflation, Real Income and Unemployment Based on Analysis of the Paper Industry and All Manufacturing," a paper prepared for the Pollution Capital Task Force of the American Paper Institute, March 1975, p. 11, estimates the cost of a net ton of new capacity to be $375.

3. See Lippke, Hughes, and Carrougher, ibid., for a discussion of these short-run influences.

Appendix 3A
Pulp Industry Impact
Estimation Model

Table 3A-1 contains the actual equations in the computer simulation model used to estimate the dynamic impact of water pollution controls on the pulp industry.

Equations (3A.1) through (3A.4) correspond to equations (3.1) to (3.4) in the text; that is, they give supply, demand, investment, and capacity in the absence of the 1972 regulations. Equation (3A.5) gives the undepreciated portion of the original capital stock and equation (3A.6) shows what part of the original industry capacity remains.

Equations (3A.7) through (3A.11) are analogous to equations (3A.1) through (3A.5) but under the force of effluent controls and a Type 1 investment assumption.

Equation (3A.12) indicates that net investment in pulp capacity is that fraction (0.961) of total investment in new capacity (total investment, $IR1$, less the capital expenditures, $OLDCLEAN$, to bring old capacity into compliance), directed toward actual pulping capability as opposed to abatement equipment.

Equation (3A.13) indicates that the capital costs of effluent abatement are the sum of water pollution control expenditures on new plant (0.039 $IR1$) plus the costs, $OLDCLEAN$, of retrofitting old capacity with pollution control devices.

Equation (3A.14) calculates the O&M costs associated with water pollution controls. The costs are cumulated as the various compliance dates ($PER2, 3, 4$) are reached.

Equation (3A.15) calculates the cash flow associated with changed price levels, $(P1 - P73) \cdot Q1$, less the operating costs of effluent control. Under Type 1 investment, there is no net increase in investment, hence, no net increase in capital outflow due to pollution control.

In Equation (3A.16), however, under the Type 2 investment assumption there is a negative cash flow associated with the capital expenditures to clean up old capacity.

Equation (3A.17) calculates the capital expenditures to bring the original capital stock into compliance with the Act. Only the undepreciated portion of the capital stock is cleaned up, hence, the $DEPREC$ adjustment. Capital expenditures are timed as described earlier.

Equations (3A.18) through (3A.24) correspond to equations (3A.8) through (3A.14) but reflect Type 2 investment behavior.

Table 3A-1

Model to Estimate the Impact of Effluent Abatement Costs on the Pulp Industry

Endogenous Variables:

C	Pulp capacity in the absence of the Act (millions of tons)
*CAPCOST1	Capital costs for effluent control (Type 1 investment)
*CAP1	Capacity (Type 1 investment)
*CASH1	Added cash flow (Type 1 investment)
DEPREC	Cumulative depreciation factor
IR	Investment (billions of dollars) in the absence of the Act
*IR1	Investment (Type 1 investment)
*NC1	Net productive capacity (Type 1 investment)
*NI1	Net investment (Type 1 investment)
OLDCAP	Remainder of original capital stock
OLDCLEAN	Capital expenditures to bring original capital stock into compliance with the Act
*OPMAIN1	Operating and maintenance costs for pollution abatement (Type 1 investment)
*P1	Price of pulp (Type 1 investment)
P73	Price of pulp in absence of the Act
Q	Pulp production in absence of the Act
*Q1	Pulp production in presence of the Act (Type 1 investment)

Equations:

$$Q = 0.96C \tag{3A.1}$$

$$\text{Log}(Q/POP) = A_0 + A_1 \cdot \log(P73) + A_2 \cdot \log(Y73/POP) \tag{3A.2}$$

$$IR = C_0 + C_1 \cdot \log(P73) \cdot C_{-1} + C_2 \cdot T \cdot C_{-1} \tag{3A.3}$$

$$C = B_0 \cdot C_{-1} + B_1 \cdot IR + B_2 \cdot IR_{-1} + B_3 \cdot IR_{-2} \tag{3A.4}$$

$$DEPREC = B_0 \cdot DEPREC_{-1} \tag{3A.5}$$

$$OLDCAP = B_0 \cdot OLDCAP_{-1} - X \tag{3A.6}$$

$$\text{Log}(Q1/POP) = A_0 + A_1 \cdot \log(P1) + A_2 \cdot \log(Y73/POP) \tag{3A.7}$$

$$Q1 = 0.96 \cdot NC1 \tag{3A.8}$$

$$IR1 = C_0 \cdot NC1_{-1} + C_1 \cdot \log(P1) \cdot NC1_{-1} + C_2 \cdot T \cdot NC1_{-1} \tag{3A.9}$$

$$CAP1 = B_0 \cdot NC1_{-1} + B_1 \cdot NI1 + B_2 \cdot NI1_{-1} + B_3 \cdot NI1_{-2} \tag{3A.10}$$

$$NC1 = CAP1 - X \tag{3A.11}$$

$$NI1 = 0.961 \cdot (IR1 - OLDCLEAN) \tag{3A.12}$$

$$CAPCOST1 = 0.039 \cdot IR1 + OLDCLEAN \tag{3A.13}$$

$$OPMAIN1 = (NC1 - OLDCAP) \cdot (PER2 \cdot OMNEW1 + PER3 \cdot OMNEW2 + PER4 \cdot OMNEW3) + OLDCAP \cdot (PER2 \cdot OM1 + PER3 \cdot OM2 + PER4 \cdot OM3) \tag{3A.14}$$

$$CASH1 = (P1 - P73) \cdot Q1 - OPMAIN1 \tag{3A.15}$$

$$CASH2 = (P2 - P73) \cdot Q2 - OPMAIN2 - OLDCLEAN \tag{3A.16}$$

$$OLDCLEAN = DEPREC \cdot (WHEN1 \cdot CLEAN1 + WHEN2 \cdot CLEAN2 + WHEN3 \cdot CLEAN3) \tag{3A.17}$$

Equations (3A.18) to (3A.24) correspond to equations (3A.8) to (3A.14) but with a modifier 2 to indicate Type 2 investment behavior.

where A_i, B_i, and C_i are coefficients (see Table 3A-2) and

$$T = \text{Time}$$
$CAPOFF$ = Capacity assumed to close due to Act
X = 0 or $DEPREC \cdot CAPOFF$ if $T = 1977$
$Y73$ = GNP extrapolated at 3.7 percent per year
POP = Population (millions) extrapolated at 1.3 percent per year
$PER2$ = Years 1977 to 1982
$PER3$ = Years 1983 to 1984
$PER4$ = Years 1985 and beyond
$OMNEW1, 2, 3$ = O & M costs associated with 1977, 1983, and 1985 standards, respectively, for new capacity only
$OM1, 2, 3$ = O & M costs associated with 1977, 1983, and 1985 standards, respectively, for old capacity only
$WHEN1, 2, 3$ = Incidence of 1977, 1983, and 1985 standards, respectively (see Table 3A-3)
$CLEAN1, 2, 3$ = Capital costs of 1977, 1983, and 1985 standards, respectively

*There exists another version of this variable with the same name, but with a modifying 2 corresponding to Type 2 investment behavior.

Note: All figures are in 1973 dollars.

Table 3A-2
Parameters of the Model: Pulp Industry

Demand Parameters	Elasticity Assumption			Capacity Parameters		Investment Parameters	
	(1) −2.33	(2) −1.0	(3) −.50				
A_0	9.45	1.77	−1.12	B_0	0.987	C_0	−159.5
A_1	−2.33	−1.00	−0.50	B_1	0.004	C_1	26.7
A_2	0.66	1.09	1.25	B_2	0.005	C_2	0.357
				B_3	0.003		

Table 3A-3
Incidence Assumptions in the Model: Pulp and Paper Industry

Year of Expenditure	Compliance Standard		
	1977	1983	1985
1974	20%		
1975	20%		
1976	60%		
1977			
1978			
1979			
1980		20%	
1981		20%	
1982		60%	
1983			40%
1984			60%
Total	100%	100%	100%

Note: Entries indicate the percentage of compliance capital costs expended in a given year to satisfy a given standard.

4

The Iron and Steel Industry

An-loh Lin
and Robert A. Leone

The manufacture of steel products involves several production stages. Each of these stages or processes generates wastewater either directly from material processing (e.g., rolling and cold finishing) or indirectly as a result of air pollution controls (e.g., wet scrubbers for sintering plants). In this chapter, we estimate the cost to the industry of federal efforts to control these sources of water pollution through the 1972 amendments to the Federal Water Pollution Control Act.[a]

Pollution Control Capital Expenditures

Table 4-1 presents estimates of the capital expenditures for water pollution control equipment the U.S. iron and steel industry will require to comply with the 1972 amendments to the Federal Water Pollution Control Act. These capital costs are both for facilities existing as of January 1973 and for projected increases in productive capacity from 1973-1983. The exhibit shows both all-inclusive cost and impending cost measures as defined in Chapter 1.

The capital costs to achieve the effluent limitation levels corresponding to the "best practicable technology" (BPT) are about $2.66 billion in 1973 dollars. To advance from BPT to abatement levels associated with the "best available technology" (BAT) will require an additional $0.43 billion, yielding a cumulative BAT capital cost of about $3.1 billion. Since abatement capital in place as of mid-1972 was about $1.06 billion in 1973 dollars, the impending cost would be about $1.6 and $2.02 billion for BPT and BAT treatment levels, respectively.[1]

In estimating the capital costs of new or expanded facilities, we used a set of published projections of the additional capacities by process necessary to increase steel production at an annual rate of 2.5 percent.[2] These

---

[a] The discussion in this chapter is limited to the costs of water pollution control. Since these costs prove to be of relatively small magnitude for an industry the size of iron and steel and since we have little evidence to support the conclusion that these costs are likely to result in major short-run dislocations in the industry, we chose not to discuss the translation of these costs into dynamic price and quantity impacts. To do so would repeat a series of arguments the essence of which is found in several other chapters.

Table 4-1

Capital Costs of Water Pollution Abatement: Iron and Steel Industry
(in millions of 1973 dollars)

	Capital Costs[a]	
	1977(BPT)	1983(BAT)
For existing facilities in place as of January 1973		
All-inclusive costs	2,656	3,084
Impending costs	1,599	2,027
For projected new and expanded capacity[b]		
All-inclusive costs	211	582
Impending costs	211	582
Total		
All-inclusive costs	2,867	3,666
Impending costs	1,810	2,609

[a]Capital costs are reported on a cumulative basis.

[b]BAT treatment standards are applied to all new and expanded facilities installed during the 1973-1977 time period.

Source: An-Loh Lin, J. Royce Ginn, and Robert A. Leone, report to the National Commission on Water Quality, August 1975, Exhibit 26.

costs are $211 million in 1973 dollars for the period 1973-1977 and $371 million for 1978-1983. In estimating the costs for this new capacity, the BAT treatment standards are àpplied as a proxy for "new source performance standards" required by the 1972 law.

Annualization of Capital Costs

In other chapters in this volume capital costs are converted into an equivalent annual cost simply by positing a rate of return (10 or 15 percent) and an expected capital life (10 to 20 years of straight-line depreciation). In this chapter, we demonstrate the use of more sophisticated procedures, which attempt to take into account important differences in federal, state, and local tax laws. These procedures yield several alternative capital recovery factors (CRF's). The differences among them illustrate important methodological points.

CRF #1

The first recovery factor simply assumes a 9 percent real rate of return and an 18-year service life for capital. No special account is taken of taxes or debt-equity ratios; and, thus, *CRF #1* corresponds, in concept, directly to the capital recovery factors used in all other chapters in this volume to annualize capital costs. *CRF #1* equals 0.114.[b]

The estimated capital life of 18 years is derived from engineering estimates.[3] The discount rate is arbitrarily assumed to be 9 percent. This is an amount slightly less than the average after-tax return to equity in U.S. manufacturing operations in recent years. Historically, returns to capital in the steel industry have been less than those in other manufacturing industries.

CRF #2

The second annualization factor attempts to account for provisions of the tax law; and, hence, requires assumptions regarding federal, state, and local tax rates on income, property, and sales.

Table 4-2 presents one set of plausible parameters which permit the computation of *CRF #2*. Several of these parameters require rather close scrutiny of state and local tax laws and tax administration for proper estimation. For example, the true rate of property taxation, p, must take into account different effective rates of property assessment as well as the nominal property tax rate.

Other input parameters vary on a firm-by-firm basis. Most noticeably, the debt constucture, d, and the real borrowing rate for capital, r, can vary substantially by firm. Unfortunately, we lack the firm-level detail necessary to take these variations into account.

The parameters in Table 4-2 reflect average values in the steel industry, and as such, need not reflect the actual numerical value of *CRF #2* experienced by any real firm. The parameters underlying *CRF #2* assume no preferential tax treatment at the federal, state, or local level for investments in pollution control equipment. The resulting *CRF #2* is 0.127, a factor only slightly higher than *CRF #1*.

To illustrate how *CRF #2* is calculated, consider a steel firm wishing to invest $1 in pollution control equipment. Assume that this $1 piece of equipment is to be replaced at the end of its service life.[c] The impact of the pollution control investment on the present value of the firm's cash flows, *dCF*, after taking all associated tax liabilities and savings into account, will be:[4]

$$dCF = \frac{-K}{1 - (1 + i)^{-n}} + d - \frac{dgr}{i} \qquad (4.1)$$

---

[b] This figure indicates that the discounted present value at 9 percent of an annual cash flow of 11.4¢ is equivalent to an initial capital outlay of $1 if the investor can earn 9 percent during the 18-year life of the equipment.

[c] We assume that the firm locates where no state investment tax credit is allowed and where state and local taxes are deductible from the federal corporate income tax base but not vice versa. This is the actual situation in those states with major steel production.

Table 4-2
Alternative Assumptions Used for Computation of Capital User Costs: Iron and Steel Industry

Parameter	CRF #2	CRF #3
Federal corporate income tax rate	40%	40%
State corporate income tax rate	7.55%	7.55%
Rate of sales tax[a]	3.97%	0.22%
Rate of property tax[a]	2.38%	0.1%
Book life of capital	18 years	18 years
Tax life of capital	15 years	5 years
Real after-tax discount rate	9%	9%
Real borrowing rate[b]	6%	3%
Debt/(equity & debt) ratio[c]	27%	27%
Investment tax credit	7%	7%
Depreciation method	sum of year's digits	straight line

State	Weight[d]	State Corporate Income Tax Rate	CRF #2 Rate of Sales Tax	CRF #2 Rate of Property Tax	CRF #3 Rate of Sales Tax	CRF #3 Rate of Property Tax
Pennsylvania	28.4%	9.5%	8%	1.25%	0 %	1.25%
Ohio	22.2	8	4.5	1.40	0	0
Indiana	19.8	5.5	0	3.58	0	0
Illinois	11.2	4	4	6.85	0	0
Michigan	9.2	7.8	4	1.47	0	0
New York	5.3	9	4.25	2.34	0	0
California	3.8	9	5.75	2.76	5.75	2.76
Average	100.0%	7.55%	3.97%	2.38%	0.22%	0.1%

[a]These average rates are weighted averages of state or local tax rates in states with major steel production.
[b]A 6 percent interest rate of tax-exempt pollution control revenue bonds is assumed.
[c]This is the average ratio observed for the steel industry in the past 5 years.
[d]Based on raw steel production in 1973.

Source: An-Loh Lin, J. Royce Ginn, and Robert A. Leone, report to the National Commission on Water Quality, August 1975, Exhibit 4.

where: K = the present value at a discount rate of i percent of the change in cash flow over an investment cycle of n years resulting from a \$1 investment financed entirely out of retained earnings: $K = 1 - c + g(s + yp) - z(1 - g)$

c = the federal investment tax credit as a proportion of the initial investment

g = the ratio of after-tax to before-tax income ($g = 1 - t^F - t^S + t^F \cdot t^S$) and assumes that the tax depreciation is the same at both the state and federal levels.

t^F = the federal corporate income tax rate

t^S = the state corporate income tax rate

s = the rate of sales taxation

y = the present value at a discount rate of i percent of the property tax basis of the investment over its service life of n years

p = the rate of property taxation

z = the present value at a discount rate of i percent of the cash flows associated with the depreciation of a \$1 investment for tax purposes

i = the real after-tax rate of return to equity

n = the physical service life of capital in years

d = the ratio of total debt financing to total invested capital

r = the real interest rate on borrowed capital

In order to fully recoup the negative discounted cash flow described in equation (4.1), the firm must be compensated by an annual gross (before-tax) income equal to

$$CRF \ \#2 = \frac{-dCF \cdot i}{g} \tag{4.2}$$

CRF #2 measures the imputed price of capital services flowing from a dollar of pollution control investment.

CRF #3

The third capital annualization factor we wish to discuss explicitly treats investments in pollution control equipment which are often subject to

Table 4-3

Percentage Reductions of the Capital User Cost Under Alternative Preferential Tax Treatments: Iron and Steel Industry
(percent)

Percent Reduction	Sales Tax Rate	Property Tax Rate	Tax Life	Borrowing Rate
25	0.9	3.3	4.3	3.2
50	1.8	6.7	9.3	6.4
75	2.7	10.0	15.1	9.6
100	3.6	13.4	25.9	12.8

Note: Computation is based on formula (4.2). Comparison is made with reference to *CRF #2* given in Table 4-2.

preferential tax treatment. These preferential tax treatments include a lower rate of (or the exemption from) sales and/or property taxes in many localities; a shorter capital life for tax purposes; and a lower interest rate associated with tax-free revenue bonds.

One set of plausible parameters which permits the calculation of *CRF #3*, using formulas (4.1) and (4.2), is reported in Table 4-2. The resulting *CRF #3* for investments in pollution abatement equipment averages 0.088 in the major steel producing states. This factor is 31 percent lower than the 0.127 CRF obtained for ordinary capital investments.

The low value of *CRF #3* is attributable primarily to the very short tax life currently allowed pollution control investments at the federal level. Table 4-3 shows the sensitivity of the annualization factor to variations in several important input parameters. The numerical value of the tax-adjusted annualization factor is actually quite sensitive to several parameters besides tax life.

For example, full property tax exemption can lower a *CRF #2* of 0.127 by over 13 percent. Similarly, interest-free money to support investments in pollution abatement equipment would lower *CRF #2* by an additional 13 percent. If capital expenditures for pollution equipment could be expensed on a current basis for tax purposes, then even in the absence of any other tax preferences, a *CRF #2* of 0.127 would drop by close to 26 percent.

The annualization factor is least sensitive to favorable sales tax provisions. A full exemption from the sales tax—in the absence of other tax preferences—lowers the *CRF #2* reported in Table 4-2 by less than 4 percent.

CRF #2 and *CRF #3* reflect average rates of taxation at the state level for those seven states representing the major portion of U.S. steel industry production. Because the tax-adjusted capital recovery factors differ by state, these variations may be the source of competitive advantages or disadvantages for specific firms in the industry.

Table 4-4

Capital User Costs for States with Major Steel Production

(in dollars per dollar investment)

States	CRF #2	CRF #3	CRF #3 ÷ CRF #2
Pennsylvania	.121	.093	.77
Ohio	.120	.084	.70
Indiana	.130	.085	.65
Illinois	.158	.085	.54
Michigan	.120	.084	.70
New York	.127	.084	.67
California	.131	.110	.84
Average	.127	.088	.69

Note: Computation is based on formula (4.2) and the information given in Table 4-2.

Table 4-4 indicates that the tax-preference adjusted *CRF #3* varies from a high of 0.110 in California to a low of 0.084 in the states of New York, Michigan, and Ohio.

The relative uniformity of the *CRF #3's* in the various steel-producing states indicates that state and local tax preferences for pollution abatement investment are not a significant source of geographical quasi-rents to individual firms in the steel industry. An impact analysis need not take these preferences into account as long as *all firms* have uniform debt-equity ratios. If, however, some firms have higher debt-equity ratios than others, then special tax provisions for pollution abatement equipment can lead to intra-industry shifts in competitive advantage. Firms that finance pollution control equipment internally will be at a relative disadvantage to firms that elect the lower-cost and tax-deductible methods of finance.

Of course, one could argue, alternatively, that firms with relatively low debt levels are better able to alter their debt structures to accommodate higher future levels of debt financing, and, hence, on the margin make better use of the special tax advantages afforded pollution control investments.

Choice of an Annualization Parameter

The above analysis suggests that variations in provisions of federal, state, and local tax laws alter the factor used to annualize the capital costs of pollution abatement equipment to the industry. Which of the several factors we have identified above should one choose in an estimation of industry impact?

The arguments in favor of *CRF #1* are fairly persuasive: It requires a minimum number of assumptions and yields a number not too dissimilar from the more complex versions of the annualization factor.[d] Further, it better reflects the real resource costs of pollution control investments to society at large.

It could also be argued that in the long run, competitive pressures from within the industry, and from products outside the industry, will compel firms to pass forward to their customers any relative cost advantages or disadvantages that result from tax laws as reflected in the tax-adjusted CRF.[e] In this case, the *CRF #2* or *CRF #3* would be neither a valid measure of the social cost of pollution abatement, nor a reliable indicator of permanent changes in the competitive balance within the industry.

The arguments in favor of choosing either *CRF #2* or *CRF #3* are also persuasive, however. For one thing, management almost surely considers tax factors when contemplating any investment, including a pollution control investment. Further, geographical disparities in competitive advantage arising from differential tax treatments are only slowly neutralized by management's ability to choose geographic location and financial structure.

The use of a tax-adjusted annualization factor is particularly relevant when losses are involved. Any losses experienced by a firm due to pollution controls decrease its net worth. The extent of the decrease depends on prevailing tax rates.

To properly account for losses, however, it is not appropriate to use the tax-adjusted annualization factor reported in equation (4.2), for this equation calculates the positive cash flow necessary to offset the negative cash flow associated with a pollution investment outlay. (These positive cash flows would result from higher prices.) If there is no positive cash flow—that is, if prices do not rise to reflect the added costs of abatement—then the entire negative cash flow must be absorbed as a loss to the firm. The loss to the firm is only in after-tax dollars, however; hence, the appropriate annualization factor to be used to calculate losses is

$$\text{loss-related } CRF = i \cdot dCF = g \cdot CRF \; \#2 \qquad \text{(or } \#3) \qquad (4.3)$$

---

[d]Clearly, the difference between an annualization factor of 0.114 and 0.088 is close to 30 percent. This difference seems less significant, however, in the light of two facts. First, a 30 percent difference in the capital cost annualization factor will lead to perhaps a 10 percent difference in total annualized costs since capital costs are approximately one-third of total annualized costs. Second, the intra-industry distribution of costs, which is fundamental to impact analysis, is largely unaltered by changes in this constant term.

[e]Of course, the higher or lower prices that result from this action will lower or raise total consumption of the taxed product. This effect, however, is only of second-order importance—particularly given the relatively small differences among the three capital recovery factors. Any interpretation of fiscal incidence is further complicated by the fact that revenues raised or foregone by the government due to one tax are presumably reflected in offsetting changes in another tax. Clearly, the net incidence of all this fiscal maneuvering is not easily disentangled.

where i is the real after-tax rate of return to equity and dCF is as described in equation (4.1).

Since g averages 55 percent, the loss-related CRF #2 is 0.55×0.127 or 0.07. The loss-related CRF #3 is 0.048.

The reason for a different loss-related cost annualization factor can be easily illustrated. Consider a firm with $1 in higher costs (all of which are deductible from income for tax purposes) and a 50 percent tax rate. To avoid a net loss, revenues must rise by $1 to cover these added costs. If costs rise by $1 but revenues remain unchanged, however, then the firm has a pretax loss of $1 and an after-tax loss of only $0.50. The appropriate annualization factor to calculate this annualized loss, therefore, depends on one's assessment of the long-run viability of the firm in question. If it can expect to offset its abatement costs with higher prices, then the annualization factor will be higher than in the case where it expects to absorb abatement costs internally.

The assumption that prices will rise in the long run to cover higher production costs is, in essence, a "going concern" assumption; that is, in the long run, any firm that is to survive in the industry must recover its full production costs.

For the purposes of this chapter, we have employed a value of 0.127 to annualize capital costs for pollution equipment. This is equivalent to CRF #2 in Table 4-4. Use of this factor implicitly assumes that there will be revenue offsets to cover the higher production costs associated with water pollution abatement; it also assumes no change in the industry's debt-equity ratio and no intra-industry variation in any of the eleven input parameters.

We chose not to employ the tax-adjusted CRF #3 on the grounds that the continued existence of these tax provisions was uncertain.

Although there are several obvious oversimplifications in the assumptions required to calculate CRF #2, we do not feel that they materially alter the nature of our conclusions.

Annualized Costs of Pollution Control

Table 4-5 shows estimates of the annualized costs of water pollution control for the steel industry. For existing facilities in place as of January 1973, the total annualized cost to achieve BPT treatment levels is about $750 million in 1973 dollars. The corresponding BAT figure is $959 million, 28 percent higher than the BPT estimate. Excluding the annual costs associated with control systems in place as of mid-1972, the BPT and BAT annualized costs will be $452 million and $661 million, respectively. Using 123 million tons of steel-mill products as a basis, the equivalent per ton annualized cost would

Table 4-5
Annualized Abatement Costs: Iron and Steel Industry
(in millions of 1973 dollars)

	Annualized Costs	
	BPT (1977)	*BAT (1983)*
For existing facilities in place as of January 1973		
All-inclusive costs	750	959
Impending costs	452	661
For projected new and expanded capacity		
All-inclusive costs	66	184
Impending costs	66	184
Total		
All-inclusive costs	816	1,143
Impending costs	518	845

Note: Costs are reported on a cumulative basis.

Source: An-Loh Lin, J. Royce Ginn, and Robert A. Leone, report to the National Commission on Water Quality, August 1975, Exhibit 26.

be $3.70 and $5.40 for BPT and BAT levels. The unit abatement cost as a percent of product price would be approximately 1.8 percent and 2.7 percent, respectively.

For new facilities projected during the 1973-1983 period, the annualized costs are about $66 million annually in 1973 dollars for net investment during the 1973-1977 time period and about $184 million annually for net investment during 1978-1983.

Unit Abatement Costs by Type of Product

Although it is difficult to assign the abatement costs we have just discussed to specific steel-mill products, it is particularly instructive to do so, for such an exercise can identify individual product lines disproportionately impacted by efforts to reduce water pollution. To permit a cost allocation to products, we first assumed that their production is an integrated process, requiring coking, iron making, steel making, casting, and other processes, depending on the type of product.

We then estimated unit abatement costs by process. Subsequently, we determined the input-output coefficients for these processes. Finally, we combined the abatement costs by process with the input-output coefficients to arrive at unit abatement costs for different finished products.

Table 4-6

Unit Water Pollution Abatement Cost by Steel Mill Product: All-Inclusive Costs

(in 1973 dollars per ton)

	BPT		BAT	
Steel Mill Product	*Unit Abatement Cost*	*As Percent of Price*	*Unit Abatement Cost*	*As Percent of Price*
Heavy structurals	3.99	2.3	5.41	3.2
Hot rolled bar & rods	4.23	2.5	5.68	3.4
Cold finished bar & rods	5.38	2.2	7.62	3.2
Wire	5.30	2.5	7.75	3.7
Seamless pipe	4.09	2.4	5.53	3.3
Welded pipe	6.38	3.0	8.10	3.8
Plate	3.56	2.1	4.96	2.9
Hot rolled sheet	4.57	2.7	6.02	3.6
Cold finished sheet	7.30	3.5	8.94	4.2
Coated products	7.73	3.5	9.40	4.3

Notes: Cost includes annualized capital costs and annual operating and maintenance costs. It includes those costs associated with in-place control systems as of mid-1972, as well as costs yet to be incurred.

Unit abatement cost for each product is the sum of all abatement costs incurred in the processes (starting with coking) of producing one unit of the product under consideration.

Computation is based on the cost coefficients and average capacities estimated for different production processes by Arthur D. Little, Inc. in a report to the American Iron and Steel Institute, May 1975, [Tables B7-B9 and Table VI-15:8].

Source: An-Loh Lin, J. Royce Ginn, and Robert A. Leone, report to the National Commission on Water Quality, August 1975, Exhibit 9.

The results are shown in Table 4-6. The abatement costs per ton of steel mill products range from $3.56 to $7.73 for BPT treatment levels and from $4.96 to $9.40 for BAT levels. As a percent of the product price, the cost varies from 2.1 percent to 3.5 percent for BPT treatment levels and from 2.9 percent to 4.3 percent for BAT treatment levels.

These unit costs are all-inclusive costs. The impending costs would be about 45 percent lower for the BPT level and about 35 percent lower for the BAT level. This means that the unit cost would vary from 1.2 percent to 1.9 percent of current product price for effluent controls associated with BPT and from 1.9 percent to 2.8 percent for BAT waste treatment requirements.

The added production costs due to water pollution controls fall rather uniformly across product lines when estimated in this manner. The differences reported here are not large and may be attributable to the uncertainties surrounding the cost assignment process.

Variations in.Unit Abatement Cost Among Plants

The unit abatement costs reported above are for finished steel mill products and are computed on the basis of average mill capacities. Actual unit costs at individual plants will certainly differ from these costs, largely because of differences in scale of individual facilities.[f]

Consider coated cold-finished sheet as an example. The unit abatement cost could represent up to 17 percent of the product price for BPT treatment levels and almost 22 percent for BAT effluent limitation levels for a plant with a 100-ton per day capacity. If the daily capacity is 15,000 tons, however, the ratio drops to 2 and 2.6 percent for BPT and BAT levels, respectively. The economies of scale are therefore very great. This suggests that there may be large intra-industry variations in the impact of the 1972 legislation due to differences in scale among plants in the industry.

We can further examine unit abatement costs as they occur among plant subcategories: nonintegrated coke plants, nonintegrated blast furnaces, integrated steel mills, steel scrap mills, and independent finishing mills. For each plant type, we first estimate the total annualized abatement costs associated with the entire plant. We then divide the total abatement costs by the estimated total revenue. The ratio measures the unit abatement cost as a percent of the composite price a plant type receives for a unit output.

Table 4-7 shows the ratios of annual abatement costs to revenues by subcategory. The ratios are about the same for integrated steel mills and steel scrap mills.[g]

We have further computed the unit abatement costs by individual plant (rather than average or "representative" plant) for the integrated steel mill subcategory and for the steel scrap mill subcategory where some individual plant data are available. The unit all-inclusive cost as a percent of the composite product price varies among integrated steel mills, ranging from 1.3 percent ($2.60/ton) to 5 percent ($10.00/ton) for BPT levels and from 2 percent ($4.00/ton) to 6.2 percent ($12.40/ton) for BAT levels. These calculations assume a price of steel of $200 per ton. These costs range from 2 percent to 4 percent of prevailing prices for the majority of the integrated mills. Those mills whose BPT cost increases are greater than 4 percent of the price of a ton of steel represent only 2.2 percent of the total raw steel capacity in this subcategory.

The unit all-inclusive abatement cost as a percent of the composite product price also varies among steel scrap mills, but to a much greater

---

[f]As we pointed out in discussing the annualization of capital costs, differences among the firms in their ability to finance pollution control investments at low interest rates will also be an important factor in determining the differential capital user costs among individual plants.

[g]Although the ratios appear low for the independent finishing mills, they are relatively high in terms of the value added which the mills receive for their products.

Table 4-7

Annual Abatement Costs per Dollar Revenue by Iron and Steel Industry Subcategory

(percent)

Subcategory	BPT	BAT
Nonintegrated coke plants	5.6	6.7
Nonintegrated blast furnaces	1.6 (2.4)	3.2 (4.8)
Integrated steel mills	2.4	3.1
Steel scrap mills	2.7	3.1
Independent finishing mills:		
Hot rolled bar & sheet	1.9 (24.4)	2.1 (27.0)
Cold rolled bar & sheet	3.3 (21.8)	4.0 (26.2)
Wire	0.9 (8.6)	5.1 (47.2)
Pipe	2.2 (15.0)	2.4 (15.9)

Notes: Costs associated with in-place control systems are included; BAT costs include BPT costs.

Total associated revenues for each subcategory are estimated as the sum for all processes of output times differential prices adjusted by input-output coefficient between two consecutive processes. Note that gross price is used for the first stage of each subcategory. Thus, all abatement costs incurred in the earlier production stages are ignored.

Figures in parentheses show the percentages of final prices contributed by value added in the subcategory mills.

Source: An-Loh Lin, J. Royce Ginn, and Robert A. Leone, report to the National Commission on Water Quality, August 1975, Exhibit 11.

degree than for the integrated steel mills, reflecting much less uniformity in sizes of individual scrap mills. These costs vary from 0.9 percent to 9 percent for BPT levels, but from 1.1 percent to 22.3 percent for BAT levels.

However, per dollar abatement costs for most of the scrap mills are quite low, due partly to the absence of pollution abatement required for coking and ironmaking processes. Only 14 scrap mills have per dollar abatement costs associated with BPT standards exceeding 3 percent of current prices. While these mills represent 12 percent of the total scrap mills, they account for only 1.7 percent of the total raw steel capacity in this subcategory. For BAT treatment levels, there are 21 mills whose per dollar abatement costs exceed 4 percent of current prices. Similarly, these mills represent 18 percent of the scrap mills but only 2.7 percent of the total raw steel capacity in scrap mills.

Technological Trends in the Steel Industry

These estimates of the cost of water pollution control in the steel industry are predicated on implicit assumptions regarding the relative stability of steel-making technology (with the exception that open hearths are assumed to be replaced by basic oxygen furnaces). There are important technological trends in the steel industry (e.g., direct reduction), which in the long run may alter these cost estimates. This concluding section briefly examines this issue.

To conceptualize the possible process implications of increasingly stringent effluent abatement requirements, it is useful to distinguish between *direct* and *indirect* production processes. Direct processes are those which bypass the intermediate product (pig iron) in the production of raw steel, including both scrap melting (in electric furnaces) and the so-called direct reduction techniques. Indirect processes are those which use blast furnaces and produce pig iron as an integral step in the metallurgical process. Changing input factor costs have differential impacts on each of these two basic systems, and both have different polluting characteristics. Consequently, the magnitude of the impact on the industry of water pollution controls depends in the long run on the industry's technological evolution.

Direct processes enjoy an advantage in water pollution abatement, due to the elimination of both coke ovens and blast furnaces. In the indirect process, coke ovens present a most complex air and water pollution control problem, while blast furnaces represent the largest single source of water use in the manufacture of steel. Also, dust emissions, which are often treated by water (wet scrubbers) in order to meet air quality standards, are substantially reduced in the direct processes. The traditional process flow of coke oven, blast furnace, and oxygen furnace emits a total of 115 pounds of dust per ingot ton, while the electric furnace emits only 2 pounds per ingot ton.

Table 4-8 presents estimates of the differential costs of abatement for the direct and indirect processes. Water pollution control costs per ton of raw steel for the indirect processes appear to be more than three times higher than those for the direct processes under BPT treatment levels, and more than four times higher under BAT levels.

Direct processes enjoy two other cost advantages over the indirect processes. The first and most important advantage is the lower capital investment necessary to install an electric furnace facility. By eliminating ore mining and processing, coke ovens, and blast furnaces, electric furnace operations can substantially reduce the capital investment necessary to produce a ton of steel-making capacity.[h] The second advantage is in the area of energy consumption. Direct processes require as little as one-sixth of the BTU's per ton of steel that the indirect processes require. Given the

---

[h] This advantage is potentially offset by the high capital requirements for scrap retrieval systems. An expansion of these systems would be required if any large increase in scrap use is to occur.

Table 4-8
Water Pollution Abatement Costs per Ton of Raw Steel for Direct and Indirect Processes
(in 1973 dollars)

Process	Total	BPT[a]		
		Steelmaking	*Ironmaking*	*Coking*
Direct	0.47	0.47		
Indirect	1.61	0.21	0.72	0.68
		BAT[a,b]		
Direct	0.61	0.61		
Indirect	2.58	0.28	1.44	0.86

[a]Unit water abatement costs for coke ovens, blast furnaces, and basic oxygen furnaces are respectively $0.21, $0.94, and $1.54 for BPT levels and $0.28, $1.88, and $1.94 for BAT levels.
[b]BAT costs are inclusive of BPT costs.

Note: The computations assume that a ton of steel requires 0.77 tons of iron and 0.45 tons of coke.

increasing scarcity of both investment capital and energy, the prospects for direct processes—and the associated savings in water pollution control costs—appear favorable. Offsetting these advantages, however, are raw material constraints, which may prevent a large-scale transfer of productive capacity to the direct processing methods in the U.S.

Summary

Not surprisingly, the iron and steel industry will incur substantial costs in complying with federal water pollution abatement requirements. The absolute magnitude of these required capital expenditures, however, is not particularly great for an industry of this size. Yet, as we continually stress throughout this volume, absolute cost levels usually impact the industry much less significantly than does the distribution of these costs among the individual member firms and plants operating within the industry. Once again, we observe substantial scale economies in wastewater treatment which, for an unchanging technology, reinforce trends toward large scale operations.

This chapter discussed in depth alternative procedures for annualizing capital costs in an industry impact study. We noted the role that tax provisions at all levels of government can play in altering the intra-industry distribution of the costs of water pollution controls.

We also noted that the steel industry has been experiencing some important trends in technology, which may be accelerated by the enforcement of strict environmental control standards. On balance, it appears that the pressures for a cleaner environment, energy conservation, and the careful husbanding of investment capital all encourage the growth of direct processes to produce steel. In the U.S., this pressure may well lead to the increased reuse of steel scrap, resulting in at least some savings in the water pollution control costs reported here.

Notes

1. For a complete discussion of all cost estimation procedures used in this chapter see An-Loh Lin, J. Royce Ginn, and Robert A. Leone, *An Economic Analysis of the Impact of the 1972 Federal Water Pollution Control Act Amendments on the U.S. Iron and Steel Industry,* report to the National Commission on Water Quality, National Bureau of Economic Research, August 1975.

2. Arthur D. Little, Inc., *Steel and the Environment,* report to the American Iron and Steel Institute, May 1975, Cambridge, Mass., Tables VI-20 and VI-21.

3. See Lin, Ginn, and Leone, op. cit., Section II.

4. For the derivation of this equation, see Michael Hanemann, *State Tax Incentives and the Annualized Cost of Pollution Control,* report to the National Commission on Water Quality, National Bureau of Economic Research, January 15, 1975.

5

The Textile Industry

Anne Hill
and Edward V. Blanchard

In 1971 the textile mill products industry accounted for 3.2 percent of the value added in manufacturing, making it the twelfth largest manufacturing industry. Although the textile industry accounts for only slightly more than 1 percent of all water used by manufacturing industries, the industry's effluent contains relatively large amounts of various pollutants, including biological oxygen demand (BOD), chemical oxygen demand (COD), oil and grease, dyes, finishing chemicals, and cleansers. Currently, very few plants recover their waste chemicals for reuse. While extensive and sophisticated waste treatment systems are in place at many larger, integrated mills, some small textile mills have no water treatment facilities.

In this chapter, we estimate the impacts of federal water pollution control legislation on this industry. In particular, we focus on the issue of plant closings caused by stringent pollution abatement requirements and the resulting geographic redistribution of plants in the industry. As in the preceding chapters, we will discuss the problems associated with measuring impact while lacking adequate data.

We begin with an examination of possible differences in the cost of compliance arising from plant managers' choices between available abatement options, and then proceed to translate these costs into impacts.

Cost Estimation Methodology

As in other studies in this volume, the costs of compliance with the Federal Water Pollution Control Act Amendments of 1972 for the textile industry are computed by extrapolating engineering estimates of compliance costs for "representative" mills across all mills in the industry.[1] These individual mill costs were aggregated to yield all-inclusive costs for the industry as described in Chapter 1 (see Table 5-1).

In our cost estimation, we discounted capital equipment at an annual rate of 10 percent over a useful life of 10 years, or (equivalently) assumed an annual "capital recovery factor" of 0.163. Since our estimates of capital costs do not include the cost of land, to derive total annual costs we include an annual cost of land of $500 per acre, which is equal to a 10 percent annual

Table 5-1
All-Inclusive Costs for the Textile Industry to Achieve 1985 Water Pollution Abatement Levels

	Mills Discharging to Surface Water		Mills Discharging to Municipal Treatment Systems	Total Subcategory Annual Costs ($ million)
	Lowest-Cost Option	Total Annual Cost ($ million)	Total Annual Cost ($ million)	
Wool and animal hair scouring	A	19.1	11.2	30.3
Wool, raw stock, top, and yarn dyeing	C	3.4	6.8	10.2
Wool and other animal hair fabric finishing	A	10.7	5.0	15.7
Woven "dry" processing mills	C	4.2	10.5	14.7
Adhesive related "dry" processing mills	C	4.3	28.8	33.1
Woven fabric finishing of cotton and cotton/synthetic blends	B	63.3	58.8	122.1
Woven fabric finishing—others	C	10.2	11.1	21.3
Knit fabric finishing of cotton and cotton/synthetic blends	C	6.0	24.9	30.9
Knit fabric finishing of 100 percent synthetics	C	12.4	17.7	30.1
Piece dyeing and printing of carpets of wool, cotton, and synthetics	B	2.8	8.2	11.0
Raw stock and yarn dyeing of cotton and synthetic fibers and yarns	C	18.9	28.8	47.7
Column total		155.3	211.8	367.1

Note: Costs include an annual cost of land of $500/acre.
Option A: The extended aeration method of wastewater treatment, including the use of lagoons.
Option B: High-rate activated-sludge process.
Option C: Spray irrigation, i.e., spraying the wastewater over agricultural or open land to let the first few inches of soil biologically degrade the wastes.

Source: M.A. Hill and E.V. Blanchard, report to the National Commission on Water Quality, 1975, Exhibit 27.

rent on a purchase price of $5000 per acre. The total annual cost for a plant is, therefore, the operating and maintenance cost, plus 0.163 times the capital cost, plus $500 times the number of acres required for that plant's wastewater treatment system.

Textile mills have several technological options for developing their own wastewater treatment systems or they may discharge their wastewater to municipal treatment systems. In 1973, approximately 75 percent of the water-using textile mills discharged their wastes to municipal systems. Since pretreating wastes before discharging to municipal systems costs significantly less than developing an entire mill treatment system, our assumption that the proportion of mills discharging to municipal systems will not increase may well overstate the final cost of water pollution abatement in this industry. On the other hand, the costs of potentially stricter state or federal controls on the wastewater streams of municipal treatment systems may well be passed back to the textile mills, who are often the major users of small-town municipal systems. This would obviously raise the cost of water pollution abatement for certain textile mills.

In addition to not knowing how many textile mills now treating their wastewater in-house will shift to using municipal treatment systems, we are not sure which in-house treatment option the mills that continue to discharge to surface water will choose. The abatement costs reported throughout this chapter assume that all mills will choose to install the lowest-cost abatement option. As indicated in Table 5-1, this lowest-cost option varies by plant, depending on its production process ("subcategory" in EPA jargon). In determining the lowest-cost option for a given mill, we considered three alternatives. Option A consists of the extended aeration method (with screening, equalization, chemical coagulation, and the use of aerated lagoons). The land requirements for this option are relatively high.

Option B, high-rate activated sludge, requires substantially less land, but has a higher capital cost than Option A and requires some in-plant process changes.

Options A and B are both "incremental"; that is, in going to each successively higher abatement level, it is generally necessary only to add additional treatment facilities and/or make additional in-plant changes. A third technology, Option C, utilizes spray irrigation to take a mill directly from no treatment to a level consistent with the 1985 goal of the Act. This technology is not incremental. Using this option to reach 1985 standards, the capital costs may be substantially less than those costs associated with meeting the same abatement level incrementally. However, since spray irrigation requires relatively large quantities of land, land generally accounts for the major portion of its cost.

In choosing an option, we assumed that all factor prices, except land, were equivalent over space. Land is a perfectly immobile production factor, required in different amounts by all water pollution abatement options. Further, the cost of land to an individual mill may vary widely, depending on the degree of congestion at the mill site and on the alternative uses to which the land might be put. Consequently, we found it useful to calculate a break-even land price to illustrate the possible sensitivity of our option choice to changes in this highly variable factor input price. Thus, while Option A may be the cost-effective option at certain land prices, as the availability and cost of land rises, there is some cost of land for each subcategory where alternative options become cost effective.

To illustrate this point, we determine the break-even land costs, above or below which an alternative to Option A treatment would be chosen. A similar calculation in Chapter 3 on the pulp and paper industry indicated that that industry had no cost-effective alternative to the land-extensive option. When comparing Option A with Option C, the break-even cost of land was that annual price per acre above which the mill will choose the extended aeration method (Option A); at any point below that price, the lower capital and operating and maintenance (O&M) costs of spray irrigation will make this option more profitable. When comparing Option A with Option B, the break-even land cost is that point above which the mill will choose Option B rather than Option A.

Given annualized capital costs, O&M costs, and the total land requirement for the spray irrigation option, Option A and Option B, the annual break-even cost of land, Pa' is determined in the following manner:

$$AC_1 + Pa' \cdot LR_1 = AC_2 + Pa' \cdot LR_2$$

$$AC_1 - AC_2 = Pa' \cdot LR_2 - Pa' \cdot LR_1$$

$$AC_1 - AC_2 = Pa'(LR_2 - LR_1)$$

so that

$$Pa' = \frac{AC_1 - AC_2}{LR_2 - LR_1}$$

where AC_1 = annual cost (O&M and capital) of the first option

LR_1 = total land required for the first option

AC_2 = annual cost (O&M and capital) of the second option

LR_2 = total land required for the second option

Pa' = annual break-even cost of land

These break-even costs are shown in Table 5-2.

Comparing the annual costs (O&M plus annualized capital costs), the

Table 5-2

Annual Break-even Costs of Land: Textile Industry (*1985 level all-inclusive costs*)

Subcategory	Annual Cost Excluding Land ($ million)			Land Required (Acres)			Annual Break-even Cost of land ($/acre)[a]	
	Option A	Option B	Option C	Option A	Option B	Option C	Option A vs. Option B	Option A vs. Option C
Wool & animal hair scouring	18.7	24.9	4.7	816	299	1,632	12,000	17,200
Wool, raw stock, top & yarn dyeing	5.1	5.1	2.8	115	32	1,198	0	2,100
Wool & other animal hair fabric finishing	10.4	10.6	9.1	469	105	4,480	500	300
Woven "dry" processing mills	16.3	17.8	3.6	1,221	220	1,374	1,500	832,200
Adhesive related "dry" processing mills	14.6	9.6	4.0	147	65	585	NA	24,200
Woven fabric finishing of cotton & cotton/synthetic blends	124.2	63.2	97.8	6,681	248	55,289	NA	500
Woven fabric finishing—others	15.4	14.5	9.9	821	31	651	NA	NA
Knit fabric finishing of cotton & cotton/synthetic blends	18.2	16.9	2.4	1,148	108	7,148	NA	2,600
Knit fabric finishing of 100 percent synthetics	18.1	13.0	8.0	927	58	8,880	NA	1,300
Piece dyeing and printing of carpets of wool, cotton, and synthetics	4.9	2.8	2.7	231	16	2,624	NA	800
Raw stock and yarn dyeing of cotton & synthetic fibers & yarns	20.5	27.3	14.3	862	85	9,234	8,800	700

NA: Not Applicable; for any cost of land, the mills in this subcategory will choose the alternative to Option A, since it has both lower total cost and a smaller land input requirement.

[a]These figures are rounded off to the nearest $100.

Source: M.A. Hill and E.V. Blanchard, report to the National Commission on Water Quality, 1975, Exhibits 25 and 26.

total land requirements, and the break-even cost of land for Options A, B, and C, it was obvious that for some mills utilizing particular production processes ("subcategories"), Option B or Option C is apparently a lower cost alternative to Option A.

As Table 5-2 indicates, the choice of treatment option in the textile industry, unlike the pulp and paper industry, is quite sensitive to the cost of land. In several cases within the textile industry, certain options dominate because they will have the lowest total cost regardless of the price of land, but in most others, the break-even land prices are at levels often observed for industrial properties. The resulting uncertainty regarding the mix of treatment options the industry will actually adopt to achieve the federally mandated effluent limitations is representative of the difficulties faced by researchers attempting to predict future industry actions with inadequate data. These limitations, in turn, raise serious questions regarding our ability to accurately assess the likely geographic consequences of water pollution control in the industry.

Plant-Level Impacts

From our estimate of costs, we turn to the problem of estimating impact by first calculating the *MCC* curve for the industry.[a] Figure 5-1 presents a set of marginal cost curves for the textile industry as a whole corresponding to 1977, 1983, and 1985 abatement levels. Costs are cumulative across levels.

From Figure 5-1, two facts are immediately clear. First, most of the industry capacity will be able to meet the 1985 (EDOP) level requirements with very low costs. More than 93 percent of capacity can be cleaned up for less than 3¢ per pound, a sum representing between 0.6 and 7.0 percent of the sales prices for various textile products.[b] Second, for any given abatement level, unit costs rise quickly for the highest cost 10 percent of the industry.

If the costs associated with the 1985 abatement level are used as a measure of maximum unit cost impact, this exhibit leads to the following conclusions.

1. Forty-eight percent of the industry will have an annualized unit cost under 1¢ per pound; 71 percent will have a cost under 2¢ per pound; 93

---

[a] For the purposes of this section, we assume all plants elect Option A. Although our own analysis indicates that, in fact, this will not be the case, the assumption greatly simplified calculations. We have also assumed the characteristics of plants, by subcategory, in the survey we utilized for this analysis are comparable to the characteristics of all plants in the industry.

[b] In 1973, the prices for textile mill products ranged from a low of $0.43 per pound for cellulosic knits to highs of $3.25 per pound for wool tops and $4.83 per pound for polyester double-knits.

Annualized Unit Cost (¢/lb.)

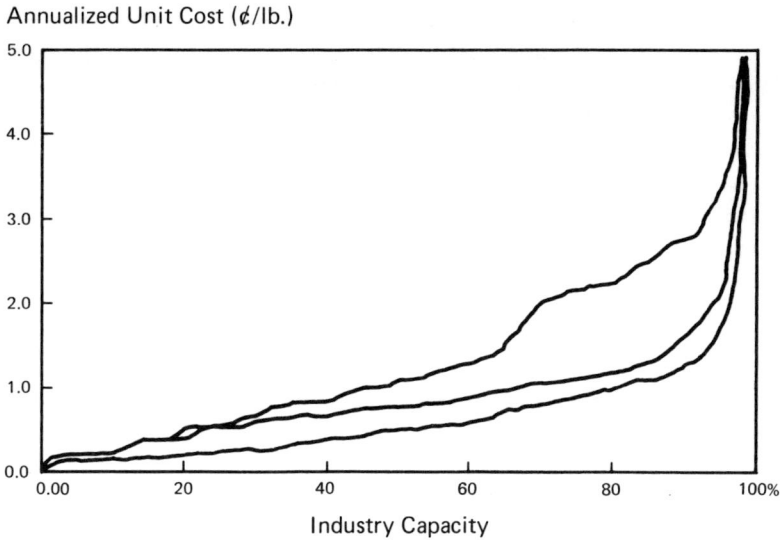

Source: M.A. Hill and E.V. Blanchard, report to the National Commission on Water Quality, 1975, Exhibit 29.

Figure 5-1. Marginal Control Cost (*MCC*) Curves: Textile Industry

Notes: Upper curve: 1985 (EDOP) Abatement levels.
                  Middle curve: 1983 (BAT) Abatement levels.
                  Lower curve: 1977 (BPT) Abatement levels.

percent will have a cost under 3¢ per pound; 99 percent will have a cost under 7¢ per pound.

2. The average cost, weighted by capacity, will be 1.7¢ per pound.
3. The last 1 percent of capacity if it is to be cleaned up would incur costs ranging from $0.07 to $3.27 per pound.

To determine more accurately the plant-level impacts of the Act, we created similar marginal cost curves for each subcategory for 1977, 1983, and 1985 abatement levels, on the assumption that the primary competition for any given mill comes from within its own subcategory. Although the shape of the marginal cost curves for most subcategories is very similar to those for the industry as a whole, cost levels vary considerably by subcategory. This variance reflects the different abatement costs associated with different mill processes. However, for each subcategory, most of the capacity (and a slightly smaller proportion of subcategory mills) is concentrated in the low-cost portions of the curve.

To determine the potential impact of the Act on the textile industry, we have defined ''potentially impacted'' mills for each subcategory as those that fall into that 10-12 percent of subcategory capacity with the highest annual costs at the 1985 abatement level. Since the unit costs of abatement rise much more rapidly for the high-cost sectors of the industry (as shown in Figure 5-1), the highest-cost 10-12 percent of capacity should include most, if not all, of those mills that face extreme additional costs as a result of the water pollution abatement requirements. The 616 mills found to be potentially impacted represent 32 percent of the mills in the industry. This implies that it is the smaller, low-volume mills in the various subcategories that face the highest abatement costs.

As discussed in Chapter 1, the competitive position of some of the potentially impacted mills may be protected by ''economic rents'' accruing to them from such sources as the servicing of small specialty markets, a general capacity shortage among a particular segment of the textile industry, or some production cost advantage relative to their competition. Thus, although some mills may appear to be potentially impacted, the unconditional association of this group with ''impact'' is quite likely to overstate the true short-run impact on prices, profits, and plant closings of the imposition of stringent water pollution controls.

To accurately determine which of the 616 high-cost potentially impacted mills can absorb or pass through to their customers the costs resulting from the Act, would require very detailed information. Lacking these data, we arbitrarily assume that any potentially impacted mill whose annualized unit cost is within 20 percent of the unit cost at the 90th percentile of capacity (which was used to identify mills as potentially impacted) will probably be able to absorb or pass on the cost increase. This is equivalent to assuming, first, that no individual segment of the textile industry will experience more than a 10 percent capacity reduction due to the Act,[c] and second, that plants with costs close to those experienced by the marginal plant are likely to stay in business despite the loss in profits caused by the added costs of effluent abatement.

We found 531 mills with costs at least 20 percent higher than the cost of the mill at the 90th percentile of its subcategory capacity. From the proportions of subcategory capacity represented by these 531 ''endangered'' mills, our best estimate is that continued existence of 4.8 percent of total industry capacity can be considered ''endangered'' as a result of the imposition of federal water pollution control standards.

The distribution of endangered mills has interesting implications for future fiber and product changes in the textile industry. Over 62 percent of the wool dyeing mills are endangered by our definition. The situation of

[c] Clearly, this first assumption is entirely arbitrary. Its only justification is that our experience in other industries, where data are available, indicates that a 10 percent capacity reduction is a large reduction, indeed. We have no information to suggest that the experience of these other industries is applicable to the textile industry.

these obviously small mills, representing together less than 9.4 percent of reported subcategory capacity, reflects the older technology of the wool processing industry with its attendant high costs of pollution abatement. The secular trend away from wool product consumption may well be accelerated by these high abatement costs.

Besides accelerating potential fiber and process changes in the industry, the differential impacts across mills may accelerate shifts in the regional distribution of the textile industry. Historically, textile mills have been small, family-owned operations located primarily in the southeastern and northeastern portions of the U.S. In recent years, large integrated textile mills have become—both through acquisition of smaller mills and through internal growth—increasingly significant in the industry, although most textile mills, even those owned by the large, financially integrated manufacturers, are still small- to medium-sized operations. With the growth of the integrated textile manufacturers, the South has increasingly dominated the textile industry; in 1967, the South accounted for 66 percent of the textile industry value added by manufacture while the Northeast accounted for only 28 percent of value added.

Various factors have contributed to this regional shift. (1) The smaller size and the age of plants in the Northeast have made them more vulnerable to technical obsolescence. (2) Textile workers' unions have tended to be much stronger in the Northeast, while a number of textile companies in the South have adamantly opposed unionization. (3) In a labor-intensive industry, the Northeast may have had higher labor costs, possibly due to the higher cost of living and to the strength of the union movement. (4) Other costs of doing business, such as taxes and energy costs, have tended to be higher in the Northeast. (5) Plants in the Northeast have also been farther from both the cotton-growing regions and many of the largest synthetic fiber plants.

The regional distribution of the potentially impacted and endangered mills may indicate the extent to which the costs of water pollution abatement will contribute to the regional redistribution of the textile industry. Because of subcategory variations in the capacity of "large" and "small" mills, we have focused on the number of potentially impacted and endangered mills in a region, rather than on the percent of total regional capacity represented by these mills. Table 5-3 arranges by region the 616 mills found by our subcategory plant-level analysis to be potentially impacted and the 531 mills found to be endangered. While the South (Region 4) has the highest number of both potentially impacted and endangered mills, these mills represent a relatively small proportion of the region's total mills, especially compared to the proportions of impacted or endangered mills in the Northeast (Regions 1, 2, and 3). This reflects the generally newer technology and larger size of southern textile mills, and implies that

Table 5-3
Textile Industry Distribution of Potentially Impacted and Endangered Mills by Region

Region	Number Potentially Impacted Mills	Potentially Impacted Mills as Percent of Regional Mills	Number of Endangered Mills	Endangered Mills as Percent of Regional Mills
1. Maine, New Hampshire, Vermont, Massachusetts, Rhode Island, Connecticut	109	36.1	104	34.4
2. New York, New Jersey, Puerto Rico, Virgin Islands	135	48.0	135	48.0
3 Pennsylvania, West Virginia, Maryland, District of Columbia, Virginia	100	39.6	85	33.3
4. North Carolina, South Carolina, Kentucky, Tennessee, Georgia, Alabama, Mississippi, Florida	231	24.8	173	18.6
5. Michigan, Wisconsin, Minnesota, Illinois, Indiana, Ohio	14	23.1	14	23.1
6. Texas, Oklahoma, Arkansas, Louisiana, New Mexico	9	28.6	9	28.6
7. Kansas, Nebraska, Iowa, Missouri	—	—	—	—
8. Colorado, Montana, Wyoming, Utah, North Dakota, South Dakota	—	—	—	—
9. California, Arizona, Nevada, Hawaii	7	17.0	4	8.3
10. Washington, Oregon, Idaho, Alaska	11	100.0	7	66.7
Total	616		531	

Source: M.A. Hill and E.V. Blanchard, report to the National Commission on Water Quality, 1975, Exhibit 46; and original materials.

the costs of water pollution abatement may contribute to hastening the secular shift in textile industry distribution from the Northeast to the South.

Our analysis of plant closings and plant-level impact has been both facilitated and limited by two implicit assumptions. First, we have assumed that the estimates of cost associated with each individual mill are, in fact, equal to the actual costs to be experienced by that mill; and, second, we have assumed that the distribution of costs in those surveyed mills for which we have data is representative of the various subcategory and industry distributions. These assumptions have allowed us to draw conclusions about the process, size, and regional characteristics of potentially impacted textile mills in the industry. Undoubtedly, unknown factors may cause the *estimates* of cost to diverge from the *actual* cost for any given mill. Although we cannot predict exactly which mills will close, we feel our cost distributions and analysis are good indicators of the types and sizes of mills most likely to suffer as a result of the costs of water pollution abatement.

Note

1. For a complete description of the methodology, including subcategory subtotals and subcategory marginal cost curves, see M.A. Hill and E.V. Blanchard, *The Economic Impact of the Federal Water Pollution Control Act Amendments of 1972 on the Textile Industry*, a report to the National Commission on Water Quality, National Bureau of Economic Research, 1975.

6

The Aluminum Industry

Wu-Lang Lee,
Robert A. Leone,
and James L. Smith

The aluminum industry has four sectors: primary and secondary aluminum smelting, bauxite refining, and fabricating.[a] Activities are typically concentrated in a very few highly integrated companies utilizing capital- and energy-intensive processes. There are only 9 bauxite refineries, 31 primary smelters, and 85 secondary smelters and refineries currently operating in the U.S. Primary smelting and refining operations are concentrated on the Gulf Coast, the Pacific Northwest, and other areas offering easy access to sources of ore and plentiful low-cost energy. Secondary smelters and refineries, on the other hand, tend to locate near major industrial and urban centers to maintain access to both scrap sources and market opportunities.

Historically, the aluminum industry has maintained a high annual growth rate (between 6.5 and 8 percent) and relatively stable prices. Stable low prices, coupled with desirable physical qualities (e.g., light weight and ductility), have contributed to the growth in consumption of this metal. Marketing representatives in the industry heavily emphasize the advantage of price stability, particularly in comparing their product to copper. Thus, some persons within the industry fear that even small price increases associated with the imposition of environmental controls will falsely signal to customers a loss of price stability. Recent erratic price behavior within the industry, however, has mooted this point.

Although aluminum is found in many compounds, its primary source is bauxite, an ore containing at least 32 percent aluminum oxide, (or alumina, as it is more commonly known). Most bauxite consumed in the U.S. is imported. As in the petroleum refining industry, the continued availability of reasonably priced raw material input is of great concern to present producers and any firm wishing to enter the industry. In recent years, governments of the resource-rich exporting countries have tightened their control over the supplies of raw materials and have drastically increased the royalties. The recent formation of the International Bauxite Association (IBA) exacerbates the situation. Thus, in the long run, the availability of bauxite may constitute a major constraint for the growth of the industry.

---

[a] Aluminum fabrication, which is essentially a dry process entailing no significant water pollution problems, lies outside the focus of our study.

On the other hand, uncertainties surrounding foreign bauxite may speed up the domestic search for alternative materials,[b] in addition to accelerating the reclamation of spent materials via recycling. Aluminum recycling looks particularly economical because secondary operations consume only one-third of the electricity the primary smelters use in producing an equivalent output.

Transportation plays a vital role in this industry because of the tonnages of raw materials required for production of finished products. The primary logistics problem in determining plant location is to find the optimum combination of low-cost energy and transportation. Although some industry observers would disagree, we believe it is unlikely that the cost of transportation will be traded off against the cost of water pollution controls in the near future. Firms may very possibly find it cheaper to transport raw materials to other countries where controls are presently either less stringent or nonexistent. We would argue that any country sophisticated enough to have an infrastructure to support a smelting mill either has or soon will have strict environmental regulations as well. Since it is virtually always cheaper to install pollution reduction equipment when a plant is first constructed, the mere threat of control is often enough to insure that the costs will be incurred.

It is worth noting that much of the wastewater generated in the production of aluminum is due to the control of air pollution—in particular, the wet scrubbing systems used to control emission of gasses. Further development of dry scrubbing systems will no doubt greatly alleviate the industry's water pollution problems. Once again we observe that the long-run impact of water pollution controls is highly dependent upon possible technological changes in the industry which may be induced by factors independent of the water pollution control problem.

Water Pollution Abatement Costs

As in all preceding chapters, the basic procedure for estimating water pollution abatement costs is to assign to individual mills costs extrapolated from "representative" or "model" plants. This approach facilitates an analysis of the distribution of costs within the industry.

Extrapolations of abatement costs are generally based on either the annual mill production or total wastewater flow. When we are aware of a unique condition in a specific plant, other criteria may also be used (e.g., flow of acid-plant blowdown).[1] Uniform application of "representative

---

[b] For example, according to a recent report, a pilot plant producing alumina from alunite has been operating at Golden, Colorado. Until recently, alunite has not been considered an economic alternative to bauxite as a source of alumina. See *Engineering and Mining Journal*, August 1974, pp. 75-76.

plant'' costs is not possible because some costs are pertinent only to specific options or processes. In such cases, we assign the costs to the relevant plants only. Likewise, if a plant requires more than one type of treatment process, the costs for this plant are reckoned as the sum of the costs of activities required to meet specific standards. In most cases, costs are estimated for three abatement levels: 1977 (BPT), 1983 (BAT), and New Source Performance Standards (NSPS).

Table 6-1 presents our estimates of impending abatement costs facing the existing aluminum smelting subcategories. The cumulative capital requirements of this segment of the industry will exceed $42 million by 1983. This corresponds to an annualized cost of nearly $24 million. When distributed over the annual output of aluminum, which exceeds 11 billion pounds, the average abatement cost per pound of aluminum falls to less than a fourth of a cent.[c] This average unit abatement cost varies only slightly between primary and secondary operations, with secondary operations having the lower costs.

Anticipated abatement expenditures of primary smelters and refiners commencing operations after January 1973 are shown in Table 6-2. These plants must conform to the New Source Performance Standards. The cost estimates in the table reflect assumed annual capacity growth at the rate of 6.5 percent. As indicated in the table, approximately $72 million in capital expenditures is required to bring the 1973-1983 incremental capacity into compliance with new source effluent standards.

Plant-Level Cost Estimates

As stressed in Chapter 1, it is the differences in abatement costs among individual plants that will cause the principal intra-industry effects.

Figure 6-1 presents the distribution of 1977 abatement costs for all existing primary and secondary aluminum smelters in the U.S. This curve reflects both annual operating and maintenance cost (O&M), and annualized capital costs, on a unit abatement basis. Since aluminum ingot produced from scrap alloys is essentially no different from ingot produced from bauxite, we combined the primary and secondary aluminum operations before forming the marginal control cost curve (*MCC*).

Despite the limited number of observations on which the curve is based, the shape is similar to *MCC* curves in other industries. In particular, the 20 percent of capacity which experiences the highest unit costs has an average cost almost four times higher than the rest of the industry. The absolute level of these costs is not great, however, and represents perhaps one half

---

[c] The market price of finished aluminum is currently (January 1976), 41¢ per pound. Hence, the average abatement cost amounts to less than 1 percent of product price.

Table 6-1

Impending Costs of Water Pollution Abatement for Existing Primary and Secondary Aluminum Smelters

(millions of 1973 dollars)

Industry Subcategory	1977 (BPT) Abatement Level				1983 (BAT) Abatement Level			
	Total Capital Cost	Annual O & M Cost	Annual Capital Cost	Total Annual Cost	Total Capital Cost	Annual O & M Cost	Annual Capital Cost	Total Annual Cost
Primary smelting	16.1	11.1	2.6	13.7	17.8	3.2	2.9	6.1
Secondary smelting	2.9	0.9	0.5	1.4	5.3	1.8	0.9	2.7
Total	19.0	12.0	3.1	15.1	23.1	5.0	3.8	8.8

Note: All costs are incremental. In computing annualized capital costs, we have assumed that a 10 percent rate of interest is charged on capital expenditures. Capital equipment is assumed to have a useful life of ten years.

Source: Wu-Lang Lee, R.A. Leone, and C.L. Smith, report to the National Commission on Water Quality, June 15, 1975, Exhibit A.

Table 6-2
Impending Costs of Water Pollution Abatement for New Sources of Aluminum Smelting
(*millions of 1973 dollars*)

		NSPS[a] for New Expansion up to 1977				NSPS[a] for New Expansion up to 1983		
	Total Capital Cost	*Annual O & M Cost*	*Annual Capital Cost*	*Total Annual Cost*	*Total Capital Cost*	*Annual O & M Cost*	*Annual Capital Cost*	*Total Annual Cost*
Primary smelting[b]	35.1	4.1	5.7	9.8	72.4	8.5	11.8	20.3

Note: This table assumes a 6.5 percent annual growth rate.

[a]New Source Performance Standards.

[b]NSPS abatement costs for secondary aluminum smelting are not available because of a lack of operating and maintenance cost data for the dry milling process.

Source: Wu-Lang Lee, R.A. Leone, and C.L. Smith, report to the National Commission on Water Quality, June 15, 1975, Exhibit 26.

Unit Abatement Cost (0.01 ¢/lb.)

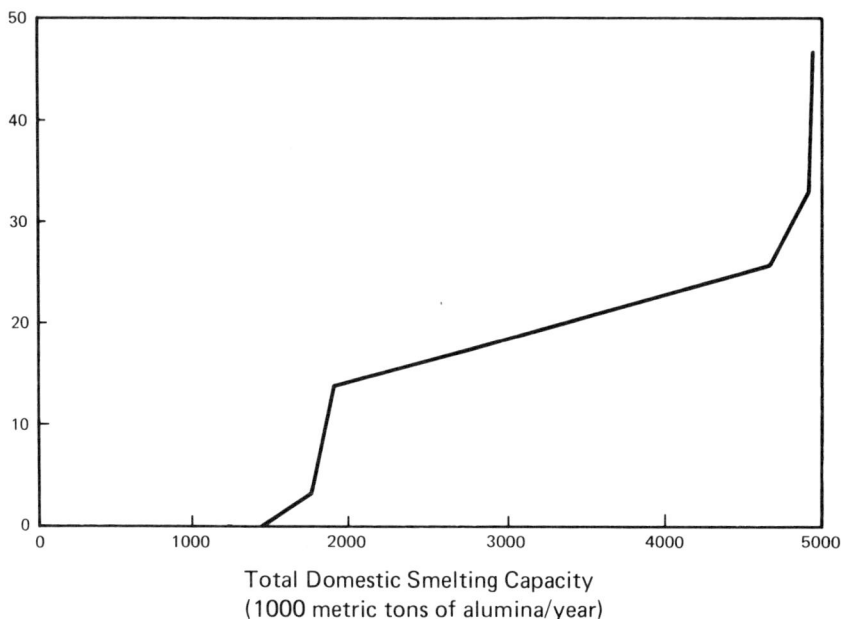

Total Domestic Smelting Capacity
(1000 metric tons of alumina/year)

Note: Figure includes impending costs only.

Source: Wu-Lang Lee, R.A. Leone, and C.L. Smith, report to the National Commission on Water Quality, June 15, 1975, Exhibit 30.

Figure 6-1. Marginal Control Cost (*MCC*) Curve for 1977 (BPT) Abatement Level: Aluminum Smelting

of one percent of the current selling price of finished aluminum. It is difficult to envision any mills actually closing in response to these relatively small cost increases.

The incidence of abatement costs among smelters is apparently less uniform than in other industries we have examined. For example, if the industry were to succeed in shifting 0.2¢, say, of the added abatement costs[d] to consumers; then 90 percent of the industry would experience profit gains ranging from 0 to 0.2¢ per pound, and the other 10 percent of the industry would experience losses ranging from 0 to 0.3¢ per pound of production. These gains and losses represent a shift in competitive advantages within the industry.

Figure 6-2 reports incremental abatement costs for the aluminum smelting subcategory corresponding to BAT abatement levels. These costs will be incurred in addition to the 1977 costs reported in Figure 6-1. Note once

[d] This cost represents marginal cost of control at approximately the 90th percentile of industry capacity.

Unit Abatement Cost (0.01 ¢/lb.)

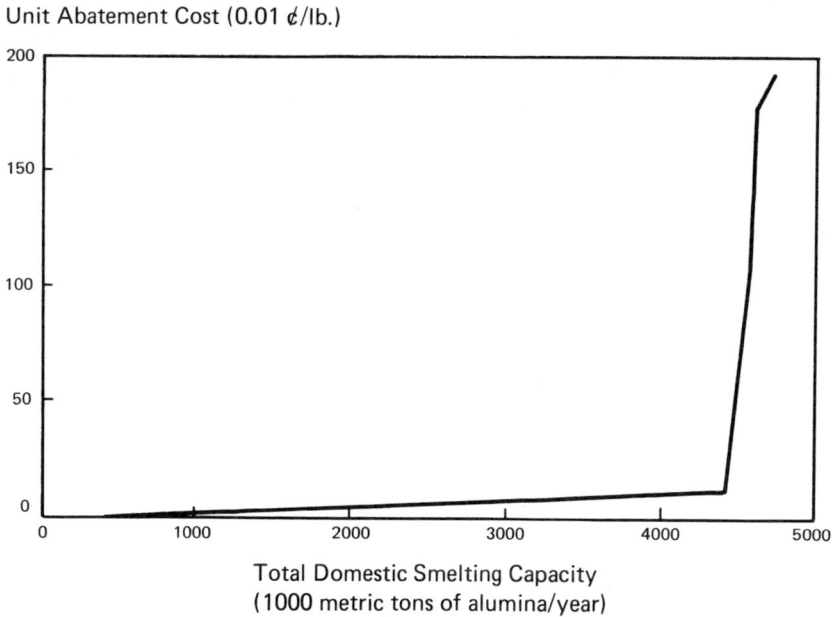

Total Domestic Smelting Capacity
(1000 metric tons of alumina/year)

Note: Figure shows impending costs to the industry to achieve 1983 abatement levels; these
costs must be incurred *in addition* to the costs necessary to achieve 1977 abatement
level.

Source: Wu-Lang Lee, R.A. Leone, and C.L. Smith, report to the National Commission on
Water Quality, June 15, 1975, Exhibit 31.

Figure 6-2. Marginal Control Cost (*MCC*) Curve for 1983 (BAT) Abatement Level: Aluminum Smelting

again that the incremental costs of more stringent controls are inconse-
quential to most of the industry. However, the mills that experience high
BPT costs also experience high BAT costs. Consequently, the stiffer 1983
standards strengthen the shift of competitive advantage caused by the BPT
requirements.

Although there is very little explicit public discussion of the role federal
pollution regulations play in shifting the competitive advantage *within* an
industry, we detect an implicit awareness of the significance of this issue.
Manifestation of this concern on the part of regulators often takes the form
of subcategories structured so as to isolate high- and low-cost plants, with
different standards applied to each. The requirements on bauxite refineries
in areas with little net rainfall (low abatement costs), for example, are quite
stringent in comparison with the standards set for the same operation in wet
areas (higher abatement costs).

Such regulatory strategies may be described as an attempt to flatten out the *MCC* curve in order to achieve the implicit equity principle of "equal pain." Understandably, various members of the industry may lack some of the regulators' enthusiasm for flattening the *MCC* curve and equalizing costs among plants. Particularly, the lowest-cost abaters can only be hurt by regulations that require them to share the burden experienced by inefficient, perhaps outmoded higher-cost abaters. We would expect that industry would like to see the *MCC* curve flattened, if at all, at a relatively low cost level. However, there is no need to dismiss avowed industry support for environmental improvement as disingenuous rhetoric on the grounds that management could never advocate a policy that increases costs. Management is in fact well aware that the upward *slope* of the *MCC* curve contributes to shifts in competitive advantage more significantly than does its absolute *level*.

Abatement Costs of Bauxite Refining

We now turn to a discussion of the abatement costs for the bauxite refining segment of the industry. Data limitations prevent us from constructing comprehensive abatement cost estimates for the bauxite refining subcategory. The best we can do is to adapt EPA estimates to our purposes.[2] The *MCC* curve in Figure 6-3 reflects the costs required for total impoundment of wastewaters from bauxite refining for the nine plants currently operating in the U.S. The abatement standards, hence costs, for the three abatement levels (BPT, BAT, and NSPS) are identical.

The following conclusions are drawn directly from the *MCC* curve:

1. Approximately 20 percent of domestic capacity is in compliance with effluent limitation guidelines and thus will incur no further costs.
2. About 53 percent of refining capacity will incur costs ranging from 0.02 to 0.044¢ per pound of aluminum.[e]
3. The remaining 27 percent of domestic supply, consisting of two refineries located on the lower Mississippi River, will be subject to abatement costs ranging from 0.45 to 0.46¢ per pound of aluminum.[f] These costs are approximately 10 to 23 times higher than those of the lower-cost abaters.
4. The industry average abatement cost will be around 0.14¢ per pound.

The impact of these abatement costs depends upon prevailing market conditions and the cost structure of the high-cost refineries. If the two Mississippi River refineries are currently operating so that prices for refined bauxite barely cover variable operating costs, imposition of abate-

---

[e] This assumes that one pound of aluminum is made from two pounds of alumina.

[f] It is of particular significance that both of these plants are owned by the same corporation.

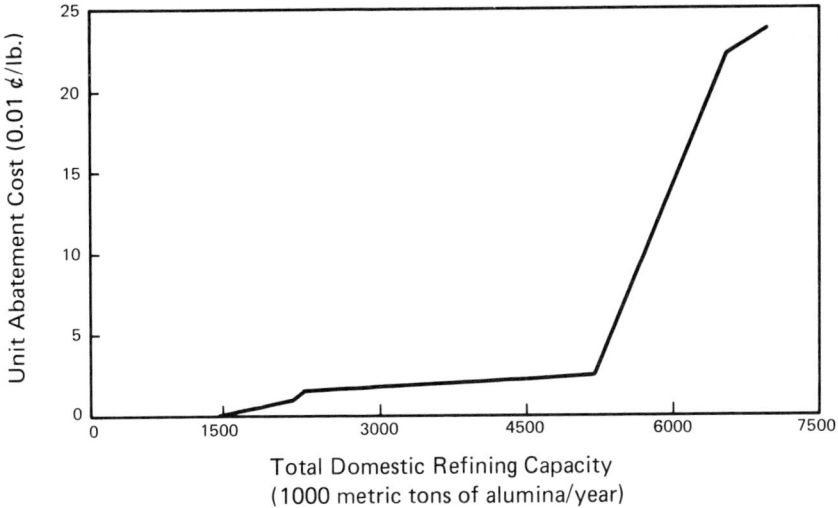

Source: "Development Document for Effluent Limitations Guidelines and New Sources
Performance Standards for the Bauxite Refining Subcategories of the Aluminum
Segment of the Nonferrous Metals Manufacturing Point Sources Category," U.S.
Environmental Protection Agency, Washington, D.C., March 1974, pp. 76-78.

Figure 6-3. Marginal Control Cost (*MCC*) Curve for All Abatement
Levels: Bauxite Refineries

ment costs would very likely cause these facilities to close. The company
which owns these plants would incur a windfall capital loss equivalent to
the present value of the stream of revenue foregone by closure less the
variable production costs at the facility.

Because bauxite refining is capital intensive, however, current operations, if profitable, probably more than cover variable operating costs. If
this is the case, and if the added costs of wastewater treatment are less than
the earned contribution of fixed costs, then these two plants will continue
to operate.

The two facilities may or may not be in a position to pass through to
customers the higher costs due to pollution controls. Consumers may be
forced to absorb at least a portion of the higher costs of pollution control in
the form of higher prices, since the rest of the industry probably cannot
make up the lost capacity should the two impacted facilities close operations. If consumer demand is strong enough to withstand a 0.45¢ per pound
increment to the price of finished aluminum,[g] then the impacted company is

[g] Abatement costs of 0.225¢ per pound of alumina processed translate approximately to costs
of 0.45¢ per pound of finished aluminum. Each pound of aluminum is derived from roughly
two pounds of alumina.

not likely to find any competitors with the capacity to shut it out of these markets. Indeed, the competing firms with low or zero-treatment-cost refineries will experience a windfall gain due to the higher prices for refined bauxite. The competitive balance in the industry will thus be affected even if marginal costs can be passed through to consumers. It should also be noted that any windfall capital loss experienced by the impacted facilities may simply offset some other cost advantage they currently possess. Perhaps some part of the competitive disadvantage of the Mississippi River facilities might be shifted back to the Jamaican suppliers of bauxite. Unfortunately, as indicated in Chapter 1, we do not fully understand how this shifting process works.

We conclude that the ability to pass costs through to consumers is of less consequence to the bauxite refining industry than is the marked shift in competitive advantages which occurs *within* the industry whether costs are passed forward or not.

Industry-Level Impact

The preceding sections of this chapter have focused on the distribution of abatement costs within the industry and have ignored any reductions in aluminum consumption that might result from higher prices for aluminum. We ignored this demand impact because the absolute levels of our cost estimates are generally low; hence, we anticipate a negligible reduction in demand. To test the reasonableness of this assumption, we explore the likely maximum long-term impact on industry demand by assuming costs are passed through entirely to the consumers of aluminum and the industry contracts as consumers respond to the higher prices by reducing consumption.

To determine reasonable upper bounds on output curtailment, we used admittedly high estimates of the price elasticities of demand. Our econometric estimate of the price elasticity of demand for aluminum ranges between -0.1 and -0.3 in the short run, and is approximately -1.7 in the long run.[3] Short-run price elasticities may be less relevant here because the time interval is insufficient to permit consumers to adjust fully to the hypothesized price fluctuations. The higher, long-run elasticity allows for the maximum substitution away from aluminum to competing materials. To allow for statistical error, we conservatively assume that the long-run elasticity of demand for aluminum is still higher: -1.8. The use of this relatively elastic measure is consistent with our attempt to measure the *maximum* demand impact of water pollution abatement.

Based on total cost pass-throughs of 0.70¢ per pound for 1977[h] and 1.7¢ per pound[i] for 1983, and a demand elasticity of -1.8, the consumption of aluminum would decline in the absence of the Act from a projected "baseline"[j] by 253,000 metric tons in 1977 and by a total of 968,000 metric tons by 1983. These reductions account for approximately 3.3 percent and 7.6 percent of projected 1977 and 1983 total U.S. capacities, respectively.[k]

Market forces are thus apparently strong enough to absorb a significant portion of abatement costs and sustain high levels of demand for aluminum, even under the extreme circumstances where increased abatement costs are entirely passed on to consumers and consumer demands are assumed to be very elastic. It is therefore reasonable to assume that our failure to include the impact of demand reductions in our previous analysis has little effect on our conclusions.

This point is reinforced by the fact that our estimate of long-run impact is extreme in several respects. First, our assumed elasticity is conservatively large. Second, competing materials also face higher costs due to pollution controls; the possibilities for substitution are thereby reduced. And third, our estimates of control costs are themselves high (assuming the validity of the engineering costs estimates). Technological change, factor substitution, and productivity gains associated with the accumulated experience with waste treatment will lower these costs, on a unit basis, over time.

Notes

1. For a detailed account of the procedures used in deriving all costs reported in this chapter see: Wu-Lang Lee, R.A. Leone, and C.L. Smith, *The Economic Impact of the Federal Water Pollution Control Act Amendments of 1972 on the Nonferrous Metals Industry*, report to the National Commission on Water Quality, National Bureau of Economic Research, June 15, 1975, Section II.

2. "Development Document for Effluent Limitations Guidelines and New Sources Performance Standards for the Bauxite Refining Subcategories of the Aluminum Segment of the Nonferrous Metals Manufacturing Point Sources Category," U.S. Environmental Protection Agency. Washington, D.C., March 1974.

3. Lee, Leone, and Smith, op. cit., pp. 17-20.

[h] 0.70¢ = 0.45¢ (highest BPT costs for bauxite refining) + 0.25¢ (highest BPT costs for primary smelting).

[i] 1.70¢ = 0.70¢ (1977 abatement level) + 1.00¢ (highest incremental BAT costs for primary smelting).

[j] Assuming the baseline consumption of aluminum in 1977 and 1983 is 8,043,300 and 12,657,200 metric tons respectively. We also assume that a 39¢ per pound price will prevail in 1977 and 1983.

[k] The combined capacity of primary and secondary aluminum operations in 1977 and 1983 is assumed to be 7.8 and 12.7 million metric tons, respectively.

7

The Metal Finishing Industry

J. Royce Ginn
and Robert A. Leone

In this chapter, we examine the impact of federal water pollution controls on the metal finishing industry. We conclude our case studies with this industry because it perhaps best illustrates the use (and limitations) of our methodological approach to impact analysis when the cost of water pollution controls is a substantial portion of an industry's total production costs.

The Metal Finishing Industry

The metal finishing industry is formally defined as all establishments classified under SIC codes 3471 and 3479: coating, engraving, and allied services (including electroplating, plating, and polishing; anodizing and coloring; and other metal finishing services not elsewhere classified). These SIC groups cover only those firms whose primary business is metal finishing; hence, formal statistical coverage of the industry is limited to the independent ("job shop") sector of the industry. Industry observers speculate that over 90 percent of the shops in the metal finishing industry are "captive"; that is, they provide metal finishing services only as part of some larger plant's production of finished goods. Our analysis tries to encompass both the "job" and hidden "captive" sectors of the metal finishing industry.

By and large, surface finishing is applied to subassemblies (parts which comprise a portion of a complete product) as an intermediate step. The finished object subsequently passes through assembly and other processes. While its costs are not major components of the final cost of a complete item, the role of metal finishing is so fundamentally important that manufacturers are unlikely to abandon its use even in the face of significant increases in the cost of finishing processes.

Aside from engraving functions and metal deposition for electrical conductivity, the industry provides finishes which it categorizes as "decorative" and "nondecorative." While these titles indicate the primary function of the finish, it is important to realize that decorative finishes are simply those that are attractive as well as resistant to abrasion, corrosion, oxidation, wear, and particle impact.

The "value of the service" provided by metal finishers may very well be much greater than its cost, but competition keeps prices close to costs. In the past, opening an independent shop required only a small amount of capital—the typical balance sheet for an independent electroplater shows less than 4 percent of sales directed toward depreciation of equipment—and except for one or two key persons (generally the owners), the labor is low skilled. Likewise, captive shops are easily set up if the independents become too expensive.

Independent shops are typically small—in 1967 almost 40 percent had fewer than five employees and 57 percent had fewer than ten employees.

Captive shops tend to be less flexible, more automated, and better financed. Since the skill level of most employees is still very low and the operating environment is hostile (because of the fumes produced by the manufacturing process), these shops are rarely viewed with pride in most plants. Cost control, quality control, and delivery scheduling are the main reasons for establishing captive shops.

Pollution Abatement Costs

The metal finishing industry extensively uses certain chemicals (e.g., cyanides, chlorides, organic acids) that are typically only incidental by-product pollutants in the operations of other industries considered major stream polluters in the U.S. Consequently, the volume of pollutants that could be emitted by this industry if wastewaters were untreated is almost beyond comprehension.

The methods currently employed to clean up metal finishing pollutants are thus, necessarily, highly sophisticated and costly. The biological treatment methods useful in so many other industries are not generally applicable here. Instead of the ordinary filters found in most waste treatment facilities, the metal finishing group must use semipermeable membranes and reverse osmosis. Frequently, the volume of chemicals required to treat the waste is as large as the volume of pollutants being treated.

According to our estimates (described in detail elsewhere)[1] the impending total cost of bringing this single industry into compliance with the 1983 standards will involve capital expenditures of $21.6 billion. The maintenance and operation of these facilities, without capital recovery, yields an *annual cost* of some $5.3 billion. These *annual* costs are almost double the one-time capital expenditures required of the steel industry to comply with the same act.

These costs entail expenditures so large that their precise impact is probably beyond our comprehension. In the preceding chapters, we have assessed the impacts of moderate increases in production costs with some degree of confidence. The impending cost changes in this industry are, indeed, so great that the techniques of incremental analysis we employ here can merely suggest the direction of the change that will occur in this industry.

The presence or absence of a likely substitute for a product can determine an industry's ability to survive dramatic production cost increases. It is difficult to imagine any readily available alternative to the metal finishing process. Virtually all types of metal require finishing (with the possible exception of stainless steel, which will suffer its own special pollution clean-up costs since it is dirtier to produce than conventional steels). Metal—because of its strength, durability, and formability—is too useful a material to be abandoned. Obviously, some metal substitutes come to mind (e.g., toasters and coffee pots made entirely of bakelite or glass) but not in sufficient numbers to threaten the existence of the metal finishing industry.

We will, therefore, continue this analysis assuming that the metal finishing industry will remain in existence. Any number of changes in the industry as we know it can make the use of metal finishing viable. Finishing processes may change dramatically, or methods for refurbishing chemicals may be developed, or metal finishing may be conducted entirely abroad, or other as yet unforeseen changes may occur.

Intra-industry Effects of Abatement Costs

Aside from competition with imports the principal determinant of the intra-industry (or competitive) effects of effluent limitations is the differential level of abatement costs among metal finishing shops serving the same market.[2] We, therefore, explicitly treat the differential incidence of pollution abatement costs among metal finishing shops.

As in the preceding chapters, our assessment of plant-level impacts begins by arraying metal finishing shops in order of unit costs of abatement due to the Act. We present these arrays—representing 1983 (BAT) and 1985 (EDOP) costs—for the entire industry in Figure 7-1.[a] We see that most of the industry is subject to a high, though fairly uniform, level of abatement costs. Some plants in the industry, however, experience extremely high costs.

As in the other industries examined, the *MCC* curve starts to turn sharply upward at a point representing 80-90 percent of industry capacity.

[a] No convenient measures of production levels are available for the metal finishing industry. A measure of "surface area finished" would be highly desirable and will probably become available when EPA regulations go into effect and encourage such record keeping. We have had to resort to using consumption of chemicals as the only available index that is even roughly correlated with production. For this index to be meaningful, we have had to assume that the impact of the regulations will encourage more efficient use of chemicals.

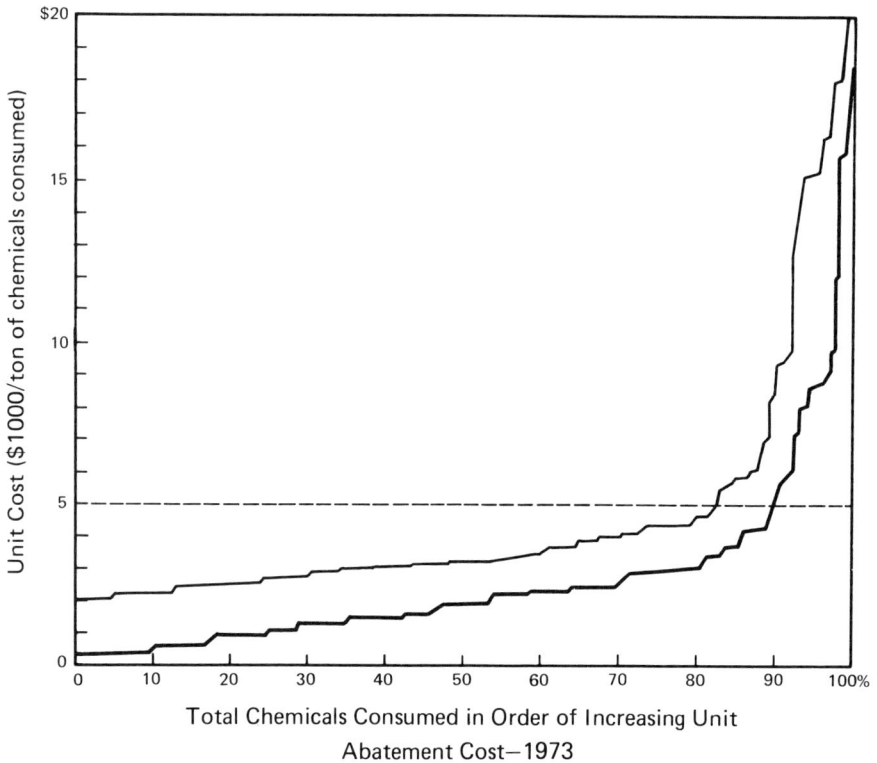

Figure 7-1. Marginal Control Cost (*MCC*) Curves by Chemicals Consumed: Metal Finishing Industry

Notes: Upper Curve: 1985 (EDOP) Abatement Level.
Lower Curve: 1983 (BAT) Abatement Level.
Costs include a proration of capital expenses (10-year life, 10% interest).

Source: J. Royce Ginn, Anne Hill, and Edward V. Blanchard, report to the National Commission on Water Quality, July 1975, Exhibit 29.

The vertical rise of the *MCC* curve in the metal finishing industry occurs when costs exceed approximately $5000 per ton of chemicals consumed for BAT levels of abatement. We would guess that metal finishing shops facing annual abatement costs greater than this $5000 level will probably be priced out of the market by the proposed effluent limitations, even if the remainder of the industry survives unchanged. In 1983, 11 percent of industry capacity will experience cost increases in excess of $5000 per ton of chemicals consumed; by 1985, 22 percent of the industry capacity will have costs in excess of $5000 per ton of chemicals used if the goal of EDOP is achieved.

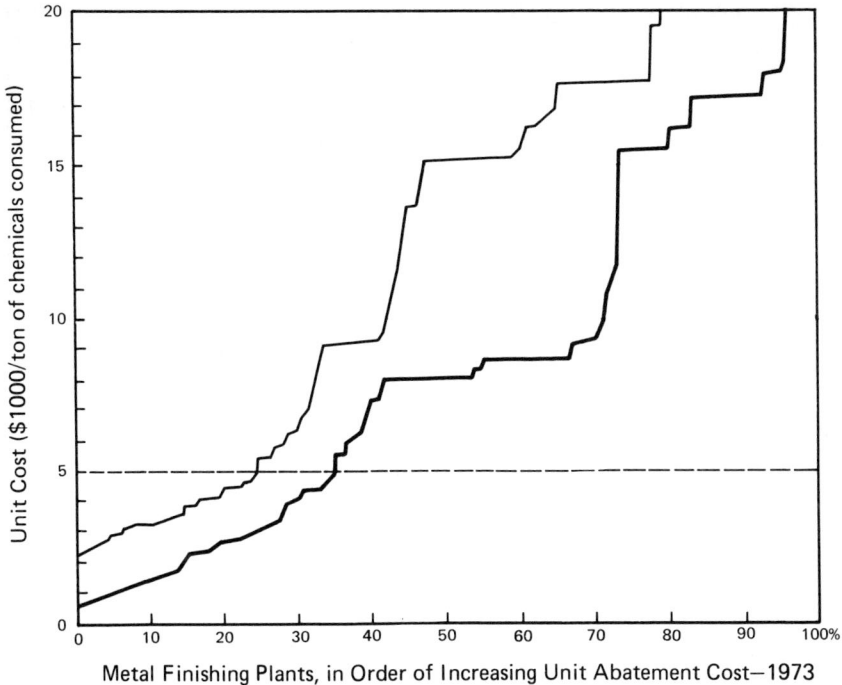

Metal Finishing Plants, in Order of Increasing Unit Abatement Cost—1973

Figure 7-2. Distribution of Unit Abatement Costs Per Ton of Chemicals Consumed: Metal Finishing Industry

Notes: Upper Curve: 1985 (EDOP) Abatement Level.
        Lower Curve: 1983 (BAT) Abatement Level.
        Costs include a proration of capital expenses (10-year life, 10% interest).

Source: J. Royce Ginn, Anne Hill, and Edward V. Blanchard, report to the National Commission of Water Quality, July 1975, Exhibit 30.

Figure 7-2 indicates that 66 percent of the plants (versus 11 percent of the capacity) have costs greater than this $5000 level for BAT treatment levels and that 77 percent would be above this level if the goal of the Act were achieved. While there may be extenuating circumstances for individual plants, it is difficult to believe that these plants can as a group survive the imposition of water pollution controls.

Of the high-cost abaters, 98 percent are captive shops in both rural and urban areas; the other 2 percent are job shops in urban areas. By the time BAT limitations have been achieved, an estimated 68 percent of all captive shops will be in this highest-cost sector of the industry. More than 40 percent of these "endangered" captive shops are small plants, performing simple operations. Although they represent only 3 percent of the total industry capacity, these plants comprise 31 percent of the capacity that is most likely to close.

The frequency with which small captive shops fall in the high-cost group of abaters raises the issue of possible misspecification of the endangered shops due to our choice of chemical consumption as the index of production. The small captive plants use only about two tons of chemicals annually, yet they are expected to have capital investments of between $60,000 and $75,000 and annual operating and maintenance (O&M) costs of $10,000-$12,000. Establishment of such shops was probably inexpensive and motivated by the need for quality control. In the face of moderately rising costs, these shops might be maintained, since the costs of handling and transporting products for outside finishing (even at nearby shops) could add significantly to the total metal finishing costs. Given large cost increases, however, their continued existence will probably depend on the size of their parent plants. Annualized pollution abatement costs in the neighborhood of $20,000-$30,000 will seem extremely large to some parent companies but negligible to others.

There are at least three limitations to the above analysis. (1) We have presented estimated rather than actual costs of pollution abatement equipment; costs of the magnitude we project here will surely encourage the development of cost-reducing technologies. (2) Our cost estimates reflect the *average* level of costs for shops of a similar type operating under similar conditions; in practice, however, costs may vary widely among shops in the same group. (3) Our method of estimating production on the basis of chemical consumption may be biased, especially across classes of plant complexity.

The most favorable interpretation of our determination of high-cost abaters is that it reflects plant-level economic impacts indicative in type and magnitude of the actual impacts likely to be felt as a result of the Act.

The Dynamic Analysis

Up to this point, we have dealt with the industry as it existed in 1973. Unless we take into account the dynamic role of new capacity expansion, however, our impact estimates will be incomplete. All existing shops compete in the long run with new finishing shops of the latest technology and design; even those shops whose markets are not seriously threatened by competition in the short run will face increasing competition as the number of metal finishing shops increases.

In order to estimate industry expansion, we first need to estimate the amount of capacity lost due to plant closures. To estimate these capacity losses we make two assumptions.

First, we assume that no firm wishing to remain in business can have annualized costs of effluent control in excess of $5000 per ton of chemicals consumed and that establishments are aware of this practical limitation. (Theoretically, we would like to have demand information to determine precisely what the market will accept in the way of higher prices for metal finishing services. Yet, given that our real uncertainties are on the supply side, this deficiency is not terribly important.)

Second, we assume that all shops that close because of water pollution abatement costs will be replaced by "complex urban job shops" in the same aggregate capacity (in tons of chemicals per year).[b] Because of the assumed difference in size, the number of new shops will be smaller than the number of shops being replaced.

The size of the complex urban job shop chosen as a replacement for plants forced to close by the Act, although arbitrary, is not a random estimate. We wanted a size small enough so that a large number of plants would be required to replace existing capacity closed by the Act, but large enough to keep unit abatement costs low enough to be competitive. The number of replacement plants for capacity lost prior to 1983—5600 plants—is more than enough to allow for one plant in each of the 3000 counties in the continential U.S. This is highly desirable because these plants are expected to service, and therefore should be accessible to, the customers of the many small shops closing throughout the U.S.

Under the first assumption, some 46,000 plants will close before the BAT regulations go into effect. Under our second assumption, the chemicals consumed in these plants represent lost production which can be replaced by the construction of 5600 new plants. Projected plant closings, between 1983 and 1985 will require the construction of an additional 5300 new plants This information is summarized in Table 7-1.

Although the substitution of complex urban job shops for existing facilities described in Table 7-1 is, of course, entirely speculative, it does show how lost capacity can be replaced (assuming market conditions allow the replacement) by an approximately equivalent amount of metal finishing capacity. Undoubtedly, the services are not exactly equivalent, especially in that the new shops will not be as convenient as the in-house shops; and their prices are likely to be higher. Also, some services available in the individually tailored small captive shops will probably not be offered by the newer plants. If such services are sufficiently valuable, they will either be

---

[b] The "standard new shop" will be a job shop providing complex metal finishing and at least one complete line of electroplating, using 60 tons per year chemicals and 100 gallons per minute rinse water in an urban area. These shops have total annualized waste treatment costs of $2020 per ton of chemicals, of which $1398 per ton represents the typical waste treatment being employed in new plants prior to the existence of the Act. These new plants will cost about $2 million to construct but will be so efficient relative to older plants that it is conceivable that they will be able to produce metal finishing at the prices very close to those that prevailed before the Act.

Table 7-1
Estimates of the Number of Plants Over Time: Metal Finishing Industry

	1983 (BAT)	1985 (EDOP)
Total plants	52,400	53,300
Original plants[a]	23,900	15,800
Replacement facilities[b]	5,600	10,900
Expansion[c]	22,900	26,600

[a]In 1973, we assume approximately 70,000 plants existed. The plants reported here are only those that survive after the Act is passed. We should note that this figure of 70,000 plants is much disputed. Discussions with persons familiar with the industry produce estimates ranging from 20,000 to 120,000 establishments.

[b]These larger-scale facilities replace those original plants that close in response to the Act.

[c]Expansion facilities are predicated on an increase in chemical consumption (our index of output) of 36.5 percent between 1973 and 1977, 7 percent between 1977 and 1983, and 5 percent between 1983 and 1985.

Source: J. Royce Ginn, Anne Hill, and Edward V. Blanchard, report to the National Commission on Water Quality, July 1975, Exhibit 37 and Section III.

offered in the newer shops or be retained in the small captive shops (which will clean up rather than close). Those services that are not sufficiently valuable to overcome the pollution abatement expenses will disappear.

Industry Expansion

In addition to construction of new facilities to replace capacity made obsolete by federal water pollution control efforts, the metal finishing industry will presumably construct additional facilities to accommodate the economy's expanding needs for finishing services.

In order to estimate the likely pollution abatement expenditures associated with expansion of metal finishing capacity we must first estimate the expected expansion. We do so using a forecast of consumer durables as an index of demand for metal finishing services, and by assuming that there is no import substitution.[3]

Given the projection of capacity expansion, and assuming new capacity will take the form of the complex urban job shop, we can determine the total costs for wastewater treatment in the new facilities. Table 7-2 shows the resulting cumulative expenditures for water pollution abatement over time.

Conclusions

We introduced this chapter by noting that techniques of incremental economic analysis cannot precisely predict the impact of the quantum produc-

Table 7-2
Impending Capital Expenditures and Annual Operating and Maintenance Costs for Water Pollution Abatement: Metal Finishing Industry
(billions of 1973 dollars)

	1983 (BAT) ($)	1985 (EDOP) ($)
Capital—total	21.5	55.1
Original plants	13.3	19.6
Replacement facilities	1.6	10.3
Expansion	6.6	25.2
Annual operating and maintenance	5.3	10.5
Original plants	3.2	4.3
Replacement facilities	0.4	1.8
Expansion	1.7	4.4

Note: See Table 7-1 for definitions of entries.

Source: J. Royce Ginn, Anne Hill, and Edward V. Blanchard, report to the National Commission on Water Quality, July 1975, Exhibit 38.

tion cost increases in the metal finishing industry associated with compliance with federal water pollution control requirements. Incremental methods simply cannot account for the magnitude of possible demand responses and supply adjustments to this new industry cost structure.

Although it is clear that our techniques cannot accurately forecast changes in this industry, it is also clear that principles of incremental analysis can suggest both the direction and magnitude of possible changes.

It is apparent, for example, that however customers react to higher metal finishing prices and however technological or organizational advances reduce abatement costs in the long run, the pressures on the highest-cost producers in this industry are real and substantial. These pressures will probably force the closing of many thousands of existing facilities. Similarly, even the low-cost technology in this industry, which seems most likely to dominate in the long run, if not supplanted by still lower treatment methods, is expensive. According to our estimates, satisfying the 1985 goal of zero discharge of pollutants would cost the existing metal finishing industry some $10 billion annually without any allowance for capital recovery even if the industry closes the highest-cost facilities and replaces them with lower cost facilities.

It is inconceivable to these authors that costs of this magnitude will actually be incurred. Although we lack the expertise to predict precisely what kinds of technical innovations, product changes, and industry reorganization will reduce these costs, it is clear to us that financial pressures of this magnitude will inevitably precipitate a dramatic—and probably unexpected—change in the way metal finishing services are delivered in the U.S. economy.

Notes

1. For details, see J. Royce Ginn, Anne Hill, and Edward V. Blanchard, *The Economic Impact of the Federal Water Pollution Control Act Amendments of 1972 on the Metal Finishing Industry*, report to the National Commission on Water Quality, National Bureau of Economic Research, July 1975.

2. For a brief treatment of the impact of imports on this industry, see Ginn, Hill, and Blanchard, op. cit., Section III, pp. 148 ff.

3. Ginn, Hill, and Blanchard, op. cit., Section III, pp. 143-144.

8

Concluding Observations

Robert A. Leone

We began this volume by noting that the current understanding of the impact of environmental control legislation on industry is rather unsophisticated. In an effort to improve the quality of this understanding, we described one methodological approach to the examination of these impacts using simple principles of microeconomic analysis. We then examined the consequences of federal efforts to reduce water pollution in six manufacturing industries within this analytical framework. There are several conclusions one can derive from these analyses which we would like to summarize here.

The Empirical Findings

Perhaps the most significant—and consistent—empirical observation in each of the industry case studies relates to the shape of the marginal cost curve for pollution abatement. Systematically the curve indicated that very large portions of most industries share common costs of pollution abatement. Typically 80 percent of an industry's capacity could be treated at essentially the same cost per unit output. The remaining 20 percent of high-cost capacity could usually be further subdivided into a high- (but not excessive) cost group and another group representing equivalent capacity but with abatement costs substantially higher than the industry average.

In every case, the 20 percent or so of high-cost capacity represented substantially more than 20 percent of the establishments in the industry, reflecting scale economies in wastewater treatment. These scale economies are themselves a potential source of long-run economic impact. As the case of the metal finishing industry illustrated rather dramatically, scale economies in pollution abatement could eliminate half of the competitors in this single industry even if total industry capacity remained unchanged.

The observation that the *MCC* curve typically begins its upward slope in the vicinity of the 80th percentile of capacity is an important empirical regularity. Principles of marginal analysis suggest that price increases due to the added costs of control should reflect the cost of control for the

117

industry's marginal producer. Since very few industries are likely to experience demand reductions of up to 20 percent in response to cost increases of perhaps 4-8 percent, it follows that in most industries we have studied, the marginal producer will experience high costs, at least in the short run. Consequently, consumers will face price increases in excess of the cost increases experienced by a large portion of the industry. The resulting *increase* in profits to firms owning those facilities that are relatively inexpensive to clean up is an important incentive, which will hasten the introduction of new capacity with low wastewater treatment costs. In the resulting long-run equilibrium, the costs of clean water to the consumer will look more like the costs on the left side of the *MCC* curve than the higher costs on the right side. There may, however, be a temporary price "blip" in the short run due to the imposition of water pollution controls.

The "blip" should occur in each of the industries we examined, with the exception of petroleum refining where the *MCC* curve only begins to rise significantly for the last 1 or 2 percent of capacity. The evidence in this industry suggested that certain individual products would likely bear the burden of these high costs.

The fact that short-run profit increases will accelerate the turnover of the industry's capital stock implies that many plants with high treatment costs will be driven out of business sooner than might otherwise have been the case. Simplistic static analysis (such as our own!) of potential plant closings may accurately state the short-run closure rate but will understate the long-run closure impact of the imposition of these environmental regulations.

The importance of the *shape* of the *MCC* curve to the determination of industry impact was particularly apparent in the study of the aluminum industry, where we noted that whether or not the higher costs of pollution abatement can be passed through to the consumer in the form of higher prices, important competitive advantages in that industry will have shifted as a result of the imposition of effluent regulations.

As we noted, the importance of the shape of the *MCC* curve can result in efforts by the regulatory agencies to "flatten out" the curve. This is typically done by applying different standards to different industry subcategories. This flattening-out exercise might well result in an "equal pain" for all affected industrial polluters, but the environmental benefits of such a flattening-out process remain unclear.

In what is a very important policy conclusion, our analyses of the petroleum and paper industries indicated that efforts to ease the impact on industry by reducing the stringency of abatement requirements are unlikely to succeed. Assuming the validity of the engineering cost data upon which we based our analyses, it appears that stricter (e.g., BAT) standards typically raise the *level* of the *MCC* curve but *rarely alter its shape*. Those plants expensive to treat to high levels of control are also relatively costly to bring into compliance with less stringent requirements (e.g., BPT).

In addition to the shape of the *MCC* curve, we identified one other factor that could contribute to temporary price "blips": capacity constraints created either by the closure of facilities unwilling to incur the costs of water treatment or by the temporary diversion of capital resources from investment in added or replacement capacity to investment in pollution abatement equipment.

We analyzed this phenomenon in some detail only in the pulp and paper industry, where we attempted to establish boundaries on this possible effect by making alternative extreme assumptions regarding the level of investment in the industry in the presence of effluent control requirements. Lacking adequate data, we identified several sets of alternative assumptions. Certain plausible combinations of these assumptions yielded simulated short-run price increases of up to 40 percent, although typically peak price increases were estimated to be in the 5-10 percent range. Clearly, sensitivity to these assumptions indicates that major advances in data collection and empirical analysis are necessary before these impacts can be estimated with some degree of confidence.

In several instances we attempted to identify possible changes in an industry's product or process mix which might reduce the costs of wastewater treatment.

In the petroleum refining industry we noted the importance of federal controls on automobile fuels in determining wastewater control costs. The requirement that automobiles use unleaded gasoline to achieve clean air standards was shown to increase the capital cost of water pollution control in this industry by some $100 million.

Opportunities to lower costs in the bleached kraft portion of the paper industry by reducing paper brightness were identified. Although it remains to be seen whether the market would accept such reductions in brightness, the exercise pointed out that opportunities for cost reduction may require innovative product marketing. This observation leads us to conclude that the task of reducing costs of pollution abatement is not one that can be limited to environmental engineers.

This conclusion was reinforced by the results of our analysis of possible process changes that might lower the costs of compliance. Internal changes in the pulp and paper industry, for example, can apparently lower the costs of effluent control significantly at some sacrifice of operating flexibility.

Other factors also contribute to the loss of operating flexibility. For example, by increasing the capital intensity of production in many industries, pollution controls can reduce operating flexibility since capital-intensive operations typically necessitate high utilization rates to insure profitability. In some extreme cases the loss of operating flexibility can be dramatic. Some biological methods for treating wastewater are low cost, but highly inflexible. During slack periods it is often necessary to literally feed these systems costly nutrients to keep the requisite biomass alive.

In the steel industry, we observed that the imposition of water pollution controls strengthens the trend to direct processes, which tend to be less polluting. In sum, we identified important trade-offs facing all types of industry management—strategists, marketing specialists, and operations managers, as well as environmental engineers—as the result of the imposition of more stringent water pollution controls.

Our analysis addressed several other less significant issues. In the steel industry, we examined the role that taxes and tax preferences play in determining the effective cost of controls to the industry. The analysis suggested that the introduction of tax parameters can modify our numerical results, but does not significantly change our conclusions.

We examined the issue of plant closings due to the imposition of water pollution controls in several of the industry studies. The lack of production cost data effectively prohibited us from making any firm conclusions in this regard. Although the source of much public discussion, our estimates of the extent of plant closings suggest only a small impact.

In several instances, we applied simple principles of cost minimization to determine which technological options management would select in given situations. The break-even land cost calculations in the textile industry, for example, demonstrated the sensitivity of option choice to this particular variable. A similar break-even exercise in the chapter on the pulp and paper industry lead to the uncomfortable conclusion that there does not exist a cost effective alternative to the land-using option. Although this conclusion may well be correct, the observation may also signal the existence of problems with the engineering cost estimates upon which our analysis is based.

In the future we would recommend using these simple break-even exercises to assess the reasonableness of engineering cost studies. When break-even calculations yield results that effectively prohibit the election of a particular treatment option, known to be in current use, either the engineering estimates of the relative costs of the options are incorrect or the specification of the "effective" alternatives needs to be reexamined.

Methodology and Data Limitations

A second set of concluding observations relates to the methodology and data limitations we faced. Our approach to impact analysis ultimately rests on the assumption that firms attempt to maximize profits in a highly competitive industrial setting. The assumption of profit maximization is probably less troublesome than the assumption of market competition. As we observed, several of the industries we examined may be considered oligopolistic in structure. Such industries do not always behave in ways that assumptions of competition would predict. Further, the very phenomenon we are studying may reduce the effectiveness of competition in the impacted industries by raising barriers to the entry of new firms. The consequences of shifting competitive structures, noticeable in the pulp and paper, textile, and metal finishing industries, are poorly understood. Although we are by no means alone in invoking an assumption of competition in an oligopolistic setting, it is a particularly questionable practice when the phenomenon being examined itself affects the validity of the underlying assumption.

A second weakness of our method is its preoccupation with "second best" analysis. Inhibited by data availability, we repeatedly were forced to make extreme alternative assumptions to permit the bounding of certain phenomenon. (In part, a "justification" for the assumption of competition is that it greatly facilitates these bounding exercises.) If the necessary data on industry production costs (as opposed to abatement costs) were available, our attention would have been more properly directed to an identification of factors that alter the shape and placement of the various industry supply curves.

This lack of data on production costs placed major limitations on our analysis, particularly in regard to the plant closure problem. Lack of data prevented us from estimating the capital losses associated with accelerated obsolescence of plant and equipment with high pollution control costs, and also severly weakened the strength of any conclusion that could be drawn regarding possible process changes due to the imposition of water pollution controls.

Lack of production cost data, coupled with both conceptual and empirical problems in estimating investment patterns, clearly reduces the overall effectiveness of the entire study.

Our poor understanding of investment behavior was perhaps most clear in the pulp and paper industry analysis. Our "rational expectations" assumption that the market will act in such a way that its decisions will be justified is a reasonable way to complete a model, but a poor way to describe an industry's response to the uncertainties inherent in any legislated (and hence reversible) changes in the industry's production function.

The methodology we employed here is particularly weak in addressing problems of short-run uncertainty, yet an increase in uncertainty may be a principal consequence of water pollution control legislation—particularly the uncertainty regarding the administrative practices of the Environmental Protection Administration.

The uncertainties facing the metal finishing industry, for example, if the abatement cost estimates reported here are to be believed, could easily cause a short-run paralysis of management decision making. It is difficult to conceive of over 50,000 entrepreneurs in a single industry simultaneously facing the decision whether or not to triple or quadruple their investment. "Rational expectations" is a meaningless concept in this context.

Our only cost data—which assume the validity of the engineering cost estimates and the reasonableness of the processes by which we extrapolate costs to individual plants—are for water pollution control costs. In every case, this is a small percentage of total production costs. Yet, even these pollution control cost data are often unsatisfactory, for the technological options we examined to achieve increasingly stringent abatement were in most cases "incremental"; that is, the technology to achieve 1983 standards was almost always an add-on to the 1977 treatment train. Spray irrigation (in the textile industry) to achieve the 1985 goal of zero discharge of pollutants was the only technology that made obsolete an option previously put in place to satisfy the requirements of a less stringent abatement level.

This almost exclusive focus on incremental technologies suggests that not all options have really been examined; it must logically follow, therefore, that the cost estimates we have reported here—even if accurate in an engineering sense—are high-side estimates of the actual costs industries will incur to comply with the law, since alternative technologies can only lower these costs.

This deficiency is most troublesome in the metal finishing industry. In the other industries, even our high-side cost estimates—again, subject to the accuracy of the engineering data—are relatively small compared to the value of the finished products. In metal finishing, however, the cost of these technologies is extremely high, and the existence of any lower-cost options could significantly alter the impact of effluent abatement on this industry.

Better data on both production costs and waste treatment options would obviously improve the quality of our analysis. Better measures of industry impact would also be possible if the individual plants for which we extrapolate costs could be aggregated by firms, thus putting in better perspective the corporate "winners and losers" due to pollution controls. Many of the firms that own plants on the low-cost end of the MCC curve, for example, also own plants in the high-cost end.

Similarly, more detail on the nature of raw material and capital goods markets would help to determine whether or not the profits or losses due to pollution control are shifted back or forward to other production factors. Many in industry contend that serious bottlenecks exist in the pollution equipment capital goods industry, for example. Hence, in the short run, they argue, the cost of pollution control is inflated as firms bid for the scarce abatement equipment. This is a possibility our current analysis cannot effectively address.

Some Generalizations

This volume has focused rather narrowly on the impact on industry of federal water pollution abatement requirements as mandated by a single piece of legislation enacted in 1972. Yet, many of the important lessons we have learned are not limited in application to this particular problem.

In fact, our impact estimation methodology applies generally to the analysis of any autonomous production cost increase industry may experience, be it an increase in energy cost, a change in local taxes, the imposition of worker safety requirements, etc. The generality of the approach we have used is at once its strength and its weakness. Its strength is that numerous empirical applications serve to prove or disprove the usefulness of this approach to impact analysis.

Its weakness is principally its failure to properly weigh several unique characteristics of cost increases, determined not in the marketplace, but by government fiat. Because environmental controls are legislatively mandated, they can be legislatively changed. This simple fact can fundamentally alter the decision-making environment in which managements operate: since the costs of pollution control are, for the most part, legislatively determined, a political versus an economic assessment of cost uncertainty is called for. Many firms do not possess the expertise to make such a political assessment and may need a period of adjustment to acquire it. Our methods do not allow for this important operational phenomenon.

A second characteristic of environmental controls is that they are to some extent subject to negotiation, often on a plant-by-plant basis. This negotiation process introduces further uncertainties (and costs) not captured in our analysis.

And third, legislative requirements are subject to the vagaries of administration. Despite passage of stiff water pollution control requirements in 1972, four years later and with the first compliance date only a year away, many industries still do not have EPA's interpretation of this law to guide them.

In sum, we have attempted to develop a methodology for estimating the impact of water pollution controls on industry, which can be generally applied to the examination of any autonomous production cost increase. To the extent we have been successful, the methods utilized here can be applied to the analysis of impact of important public policies, including other forms of environmental control, safety regulations, and the like.

Index

Index

127

115, 118; derived, 28, 30; elasticity, 4, 5, 26, 28, 30, 31-32, 50, 54, 58, 104-105; for refinery services, 27, 28, 30, 31, 33, 39, 41; long-run, 4; schedules, 2, 3f, 5-6, 7f, 13, 14f, 16f, 17, 18, 19, 29, 48, 49-50, 54, 55
Depreciation, 23, 26, 28, 36, 53, 55, 68, 70t, 71, 108
Desulphurization, 39, 40
Displacement of capital. *See* Capital, diversion of

Economies of scale, 11, 22, 78, 81, 117; municipal treatment, 37-38
Endangered plants, 36, 90-91, 92t, 111-112. *See also* Potentially impacted plants
Equal pain principle, 102, 118
End of discharge of pollutants. *See* Federal Water Pollution Control Act, Amendments of 1972, 1985 goal
Energy cost. *See* Cost of pollution control, energy
Equilibrium. *See* Impact, long-run, short-run

Factor: prices, 13, 45-46, 80, 85-86; substitution, 4, 59, 105
Federal Water Pollution Control Act, Amendments of 1972: compliance with, xiii, xv, 9, 12, 43, 67, 83, 97, 115; enforcement, xiv, 54; 1977 standards, xiii, 23, 31, 46, 67, 75, 76, 77, 78, 79, 80, 81t, 88, 89, 97, 98t, 100f, 101, 102, 105n, 119, 122; 1983 standards, xiii, 23, 31, 67, 68, 75, 76, 77, 78, 79, 80, 81t, 88, 89, 97, 98t, 100, 101, 102, 105n, 108, 109, 110, 111, 113, 114t, 115t, 118, 122; 1985 goal, xiii, 23, 31, 38, 85, 88, 89, 90, 109, 110, 111, 114t, 115t

Geographic impact. *See* Impact, regional
Grassroots facilities. *See* Cost of pollution control, new capacity

Impact: on capacity, xv, 5, 6, 7, 11, 19, 26, 27, 31, 32, 49, 50, 54-55, 59f, 90, 112-113, 119; dynamic, 6-8, 19, 35, 48, 49, 112-114; equation with costs, xiv, xv, 6, 8, 19; intra-

industry, 8-11, 72-73, 74n, 81, 97, 100-102, 104, 109-112, 118, 121; on investment, xv, 52, 54, 69, 119; long-run, xv, 4, 6-7, 19, 34, 36, 96, 104-105, 117, 118; on output, 3-4, 5, 6, 8, 13-17, 19, 26, 31, 32, 48; on prices, 2-3, 4, 5, 6, 8, 9, 13-17, 19, 25-26, 31, 32, 34-35, 48, 52, 54, 58, 59, 75, 90, 103-104, 113, 115n, 118, 119; on profits, 4, 9-10, 15-17, 25-26, 31, 45, 48, 53, 90, 100, 118; plant-level, 21, 33-34, 36, 88-93, 109, 112; public understanding of, xiii, 101; regional, xv, 10, 34-37, 74, 83, 88, 91-93; short-run, xv, 3, 4, 5, 6, 7, 19, 26, 34, 35, 36, 43, 54, 58, 60, 90, 118, 119, 121. *See also* Marginal control cost curve, plant closings; Cost of Pollution control, uneven distribution of
Integrated steel mills, 78-79
Internal changes. *See* Process changes
Iron and steel industry, xv, 4, 13, 67-82, 120

Land Cost. *See* Cost of pollution control, land
Lead-free gasoline, 40, 119
Long-run analysis. *See* Impact, long-run
Losses, 9, 10, 14f, 15-16, 17, 25, 55, 74, 100, 103, 104, 121, 122. *See also* Capital recovery factor

Marginal control cost curve, 8, 9, 10f, 13, 14f, 15, 16f, 17, 23, 31, 44, 45f, 46, 47f, 48, 55, 88-89, 93, 97, 100f, 101f, 102, 103f, 109-110, 117, 118, 119; construction of, 8-9
Maximum effects, output, 6, 32, 33. *See also* Impact, on output
Metal finishing industry, xv-xvi, 4, 107-116, 117, 121, 122; captive shops, 107, 108, 111-112, 114; complex urban job shops, 113-114, job shops, 107, 108, 111, 113
Methodology, xi, xv, xvi, 1-19, 21, 43-45, 93, 120-123; assumptions, 1-2; choice of, xiv-xv; criteria, 1; practicality of, 15; scope of analysis, xiv, 1, 2, 6

About the Editor

Robert A. Leone received the Ph.D. in economics from Yale University in 1971 where he served as lecturer in the Department of Economics and Research Associate in the Institution for Social and Policy Studies. He is presently an Assistant Professor of Business Administration at the Harvard Business School and Research Associate at the National Bureau of Economic Research.

Related Lexington Books

Brenner, Michael J., *The Political Economy of America's Environmental Dilemma*, 208 pp., 1973

Miernyk, William H. and Sears, John T., *Air Pollution Abatement and Regional Economic Development*, 224 pp., 1974

Ray, Marvin E., *The Environmental Crisis and Corporate Debt Policy*, 128 pp., 1974

Van Tassel, Alfred J., *Environmental Side Effects of Rising Industrial Output*, 576 pp., 1970